Numerical and Symbolic Computation

Numerical and Symbolic Computation

Developments and Applications

Special Issue Editors

Maria Amélia Ramos Loja
Joaquim Infante Barbosa

MDPI • Basel • Beijing • Wuhan • Barcelona • Belgrade • Manchester • Tokyo • Cluj • Tianjin

Special Issue Editors
Maria Amélia Ramos Loja
CIMOSM, ISEL, Instituto Superior de Engenharia de Lisboa
Portugal

Joaquim Infante Barbosa
IDMEC, IST, Instituto Superior Técnico, Universidade de Lisboa
Portugal

Editorial Office
MDPI
St. Alban-Anlage 66
4052 Basel, Switzerland

This is a reprint of articles from the Special Issue published online in the open access journal *Mathematical and Computational Applications* (ISSN 2297-8747) (available at: https://www.mdpi.com/journal/mca/special_issues/SYMCOMP2019).

For citation purposes, cite each article independently as indicated on the article page online and as indicated below:

LastName, A.A.; LastName, B.B.; LastName, C.C. Article Title. *Journal Name* **Year**, *Article Number*, Page Range.

ISBN 978-3-03936-952-2 (Hbk)
ISBN 978-3-03936-953-9 (PDF)

Contents

About the Special Issue Editors

Maria Amélia Ramos Loja is presently an Adjunct Professor in the Mechanical Engineering Department of the Engineering Institute of Lisbon (ISEL, IPL), an invited Associate Professor of the Physics Department of the University of Évora (UEvora) and a Senior Researcher of the Mechanical Engineering Institute (IDMEC, IST). Her academic background integrates a BSc with honors in Marine Engineering from the Portuguese Nautical School and a BSc in Computer Science. Her MSc and Ph.D. degrees in Mechanical Engineering were conferred by the IST, University of Lisbon and the Habilitation in Mechatronic Engineering by the University of Évora. Her principle areas of interest include the scientific areas of Computational Solids Mechanics, Composite Materials, Smart Materials, Optimization, and Reverse Engineering. She is Chairperson of the ECCOMAS thematic series of conferences SYMCOMP (International Conference on Numerical and Symbolic Computation: Developments and Applications) and she coordinates the Research Centre on Modelling and Optimization of Multifunctional Systems (CIMOSM, ISEL). Since 2017 she has been invited by European Commission Research Agencies to evaluate project proposals in different subjects related to her competences.

Joaquim Infante Barbosa is presently a Jubilated Full Professor for the Mechanical Engineering Department of the Engineering Institute of Lisbon (ISEL, IPL, Polytechnic Institute of Lisbon) and Senior Researcher of the Mechanical Engineering Institute (IDMEC, IST, University of Lisbon). His academic background integrates a degree in Mechanical Engineering by IST, University of Lisbon. His MSc and Ph.D. degrees in Mechanical Engineering were conferred by IST, University of Lisbon and the Habilitation in Structural Mechanics by the University of Évora. His major areas of interest include the scientific areas of Computational Mechanics, Structural Optimization, Composite Materials and Vibration Suppression, among others. He is a researcher on several Portuguese and European projects and a reviewer of various engineering journals. He is a member of the organizing committee of the ECCOMAS thematic series of conferences SYMCOMP (International Conference on Numerical and Symbolic Computation: Developments and Applications) and a senior member of APMTAC—Portuguese Association of Theoretical, Applied and Computational Mechanics.

Mathematical and Computational Applications

Editorial

Preface to *Numerical and Symbolic Computation: Developments and Applications—2019*

Maria Amélia R. Loja [1,2,3,*] **and Joaquim I. Barbosa** [1,3]

1   CIMOSM, ISEL, Centro de Investigação em Modelação e Optimização de Sistemas Multifuncionais, 1959-007 Lisboa, Portugal; joaquim.barbosa@tecnico.ulisboa.pt
2   Escola de Ciência e Tecnologia, Universidade de Évora, 7000-671 Évora, Portugal
3   IDMEC, IST—Instituto Superior Técnico, Universidade de Lisboa, 1049-001 Lisboa, Portugal
*   Correspondence: amelia.loja@isel.pt

Received: 11 May 2020; Accepted: 11 May 2020; Published: 11 May 2020

This book constitutes the printed edition of the Special Issue *Numerical and Symbolic Computation: Developments and Applications—2019*, published by *Mathematical and Computational Applications* (MCA) and comprises a collection of articles related to works presented at the 4th International Conference in Numerical and Symbolic Computation—SYMCOMP 2019—that took place in Porto, Portugal, from April 11th to April 12th 2019.

This conference series has a multidisciplinary character and brings together researchers from very different scientific areas, aiming at sharing different experiences, in a cross-fertilization perspective. Therefore, the articles contained in this book, although sharing a common characteristic related to the use of numerical and/or symbolic methods and computational approaches, also present an overview of their use in a transversal way to science and engineering fields.

In the first contribution *Bridging Symbolic Computation and Economics: A Dynamic and Interactive Tool to Analyze the Price Elasticity of Supply*, from Andraz et al. [1], the authors propose a new dynamic and interactive tool, created in the computer algebra system Mathematica and available in the Computable Document Format. This tool can be used as an active learning tool to promote better student activity and engagement in the learning process, among students enrolled in socio-economic programs.

The second article of the book is authored by Escobar et al. [2] and has the title *The Invariant Two-Parameter Function of Algebras $\overline{\psi}$.* In this article, it is proven that the five-dimensional classical-mechanical model built upon certain types of five-dimensional Lie algebras cannot be obtained as a limit process of a quantum-mechanical model based on a fifth Heisenberg algebra. Other applications to physical problems are also addressed.

Gavina et al. [3], in their article *Solving Nonholonomic Systems with the Tau Method*, propose a numerical procedure based on the spectral tau method to solve nonholonomic systems. The Lanczos' spectral tau method is used to obtain an approximate solution to these nonholonomic problems exploiting the tau toolbox software library, adding to the ease of use characteristics and providing accurate results.

The contribution of Matos and Rodrigues [4], *Almost Exact Computation of Eigenvalues in Approximate Differential Problems*, addresses differential eigenvalue problems that arise in many fields of Mathematics and Physics. These authors present a method for eigenvalues computation following the Tau method philosophy and using Tau Toolbox tools, wherein the eigenvalue differential problem is translated into an algebraic approximated eigenvalues problem, after which by making use of symbolic computations, they arrive at the exact polynomial expression of the determinant of the algebraic problem matrix, allowing us to get high accuracy approximations of differential eigenvalues.

In a different area, Monteiro et al. [5], through their article *Factors for Marketing Innovation in Portuguese Firms CIS 2014*, aim at understanding which factors influence marketing innovation and also aim to establish a business profile of firms that innovate or do not in marketing. These authors used

Math. Comput. Appl. **2020**, *25*, 28

multivariate statistical techniques, such as, multiple linear regression (with the Marketing Innovation Index as dependent variable) and discriminant analysis where the dependent variable is a dummy variable, indicating if the firm innovates or not in marketing.

The sixth article *Numerical Optimal Control of HIV Transmission in Octave/MATLAB*, from to Campos et al. [6], provides a GNU Octave/MATLAB code for the simulation of mathematical models described by ordinary differential equations and for the solution of optimal control problems through Pontryagin's maximum principle. A control function is introduced into the normalized HIV model and an optimal control problem is formulated, where the goal is to find the optimal HIV prevention strategy that maximizes the fraction of uninfected HIV individuals with the least amount of new HIV infections and cost associated with the control measures.

The contribution of Rodrigues [7] entitled *Isogeometric Analysis for Fluid Shear Stress in Cancer Cells* constitutes the seventh and last paper of this book. In this article, the author considers the modelling of a cancer cell using non-uniform rational b-splines (NURBS) and uses isogeometric analysis to model the fluid-generated forces that tumor cells are exposed to, in the vascular and tumor microenvironments, during the metastatic process. The aim of the article is focused on the geometrical sensitivities to the shear stress exhibition of the cell membrane.

At this point, as editors of this book, we would like to express our deep gratitude for the opportunity to publish with MDPI. This acknowledgment is deservedly extensive to the MCA Editorial Office and more particularly to Mr. Everett Zhu, who has permanently supported us in this process. It was a great pleasure to work in such conditions. We look forward to collaborating with MCA in the future.

Conflicts of Interest: The authors declare no conflict of interest.

References

1. Andraz, J.M.; Candeias, R.; Conceição, A.C. Bridging Symbolic Computation and Economics: A Dynamic and Interactive Tool to Analyze the Price Elasticity of Supply. *Math. Comput. Appl.* **2019**, *24*, 87. [CrossRef]
2. Escobar, J.M.; Núñez-Valdés, J.; Pérez-Fernández, P. The Invariant Two-Parameter Function of Algebras ψ. *Math. Comput. Appl.* **2019**, *24*, 89. [CrossRef]
3. Gavina, A.; Matos, J.M.A.; Vasconcelos, P.B. Solving Nonholonomic Systems with the Tau Method. *Math. Comput. Appl.* **2019**, *24*, 91. [CrossRef]
4. Matos, J.M.A.; Rodrigues, M.J. Almost Exact Computation of Eigenvalues in Approximate Differential Problems. *Math. Comput. Appl.* **2019**, *24*, 96. [CrossRef]
5. Monteiro, P.; Correia, A.; Braga, V. Factors for Marketing Innovation in Portuguese Firms CIS 2014. *Math. Comput. Appl.* **2019**, *24*, 99. [CrossRef]
6. Campos, C.; Silva, C.J.; Torres, D.F.M. Numerical Optimal Control of HIV Transmission in Octave/MATLAB. *Math. Comput. Appl.* **2020**, *25*, 1. [CrossRef]
7. Rodrigues, J.A. Isogeometric Analysis for Fluid Shear Stress in Cancer Cells. *Math. Comput. Appl.* **2020**, *25*, 19. [CrossRef]

Mathematical and Computational Applications

Article

Bridging Symbolic Computation and Economics: A Dynamic and Interactive Tool to Analyze the Price Elasticity of Supply

Jorge M. Andraz [1,2], Renato Candeias [2] and Ana C. Conceição [3,*]

[1] Center for Advanced Studies in Management and Economics (CEFAGE), Universidade do Algarve, Campus de Gambelas, 8005-139 Faro, Portugal; jandraz@ualg.pt
[2] Faculdade de Economia, Universidade do Algarve, Campus de Gambelas, 8005-139 Faro, Portugal; rakinus@outlook.pt
[3] Center for Functional Analysis, Linear Structures and Applications (CEAFEL), Faculdade de Ciências e Tecnologia, Universidade do Algarve, Campus de Gambelas, 8005-139 Faro, Portugal
* Correspondence: aconcei@ualg.pt; Tel.: +351-289800900

Received: 1 August 2019; Accepted: 9 October 2019; Published: 10 October 2019

Abstract: It is not possible to achieve the objectives and skills of a program in economics, at the secondary and undergraduate levels, without resorting to graphic illustrations. In this way, the use of educational software has been increasingly recognized as a useful tool to promote students' motivation to deal with, and understand, new economic concepts. Current digital technology allows students to work with a large number and variety of graphics in an interactive way, complementing the theoretical results and the so often used paper and pencil calculations. The computer algebra system *Mathematica* is a very powerful software that allows the implementation of many interactive visual applications. Thanks to the symbolic and numerical capabilities of *Mathematica*, these applications allow the user to interact with the graphical and analytical information in real time. However, *Mathematica* is a commercially distributed application which makes it difficult for teachers and students to access. The main goal of this paper is to present a new dynamic and interactive tool, created with *Mathematica* and available in the Computable Document Format. This format allows anyone with a computer to use, at no cost, the PES(Linear)-Tool, even without an active Wolfram *Mathematica* license. The PES(Linear)-Tool can be used as an active learning tool to promote better student activity and engagement in the learning process, among students enrolled in socio-economic programs. This tool is very intuitive to use which makes it suitable for less experienced users.

Keywords: symbolic computation; dynamic and interactive tool; socio-economic sciences; F-Tool concept; PES(Linear)-Tool; Wolfram *Mathematica*; computable document format

---

1. Introduction

The use of educational software has been increasingly recognized as a useful tool to promote students' motivation to deal with, and understand, new concepts in different study fields (see, for instance, [1–12]). In fact, educational software tools have a great potential of applicability, particularly at the university level, where the knowledge of various areas by different careers is required [8]. Current digital technology allows students to work with a large number and variety of graphics, in an interactive way, complementing the theoretical results and the so often used paper and pencil calculations. Obviously, calculations with this kind of support do not replace paper and pencil calculations, and they should be properly combined with other methods of calculation, including mental calculation. Some studies conclude that students using computer algebra systems are at least as good in "pencil and paper" skills as their traditional counterparts [13]. This aspect is not of minor relevance. Although the "pencil and paper" work can be done by computers, students should learn how calculations are made and also should learn how the computer algebra systems work [14] (we thank an anonymous referee for this observation). Also, the use of technology in the classroom can lead to advances in conceptualization, contributing thereby to students' engagements and motivation [15]. According to [16], one of the reasons for students to use computer algebra systems is their belief that these tools help their understanding of new concepts.

The computer algebra system *Mathematica*, conceived by Stephen Wolfram, and developed by Wolfram Research, is a very powerful software that allows the implementation of many interactive visual applications. Thanks to the symbolic and numerical capabilities of Mathematica, these applications are eminently dynamic tools, where the user can interact with the graphical and analytical information in real time. More importantly, the graphics are taken out of the textbook and they are placed under the user's control, where the user can manipulate, investigate, and explore their characteristics. Students who have used *Mathematica* for at least one year identified this kind of visualization as one of the significant benefits they found with the use of *Mathematica* [16].

Graphics are always helpful in the learning process, but [16] states that it makes a difference whether the students' interaction with graphic visualization is active or passive. As reported by [17], academics in higher education institutions should not only worry about the contents, but also give attention to the learning environment as they face students with different motivations and different levels of involvement. Such differences will likely affect the teaching and learning process. Moreover, teachers can expect that, in any classroom, some students prefer to be receivers (observers or listeners), while others prefer to be active participants. In fact, there are students with a more active attitude, who, even in a more traditional class, theorize, apply and relate, and there are those who exhibit more passive behavior. Clearly, these students require different orientation and teaching methods so that they are able to fully engage in the classroom activities as agents of a truly active learning process. This type of learning denotes a style of teaching that provides opportunities for students to talk, to listen to, and to reflect on what they have learned, as they participate in a variety of learning activities [18,19]. We should note that teachers who employ active learning strategies in their classrooms are unlikely to please all students all the time [20], but neither is a teacher who relies regularly on traditional lectures. The active learning also aims to improve the students' performance and develop the skills they need, for example, to obtain a better grade in a specific curricular unit [19]. In many cases, active learning can be employed without increased costs and with only a modest change in current teaching practices with a reduced risk and a high return [20]. Unfortunately, there are gaps between teaching and learning, between teaching and testing, and between educational research and practice in higher education institutions [21]. A serious gap also exists between how faculty members typically teach (i.e., relying largely on the "lecture method") and how they know they should teach (i.e., employing active learning strategies to develop intellectual skills, and to shape personal attitudes and values). Moreover, teachers see few incentives to change mainly because the use of educational software in classrooms is time-consuming. In fact, any faculty member who has ever attempted to lead a true one hour class discussion, in which students talk and respond to one another, knows how difficult it is to have control over the discussion.

Notwithstanding the above, the importance of using educational software in mathematics, as an efficient tool to help students grasp with hard-to-understand concepts and to more quickly gain a deeper understanding of the materials being taught firsthand, is acknowledged (see, for instance, [1,2,10,16,22]) and thereby such software can help to promote an active learning environment inside the classroom.

Although it is recognized that some economic concepts can be more easily understood when the students work with a large number and variety of graphics in an interactive way, with the support of the appropriate technology, the use of computer algebra systems is rare and under-studied in economics education (we thank an anonymous referee for this observation). In fact, the use of educational software in economics has been limited to some specific economic concepts (see, for instance, [23–25]). According to [23,24], there are automatic algebraic simplifiers, but simplicity is often in the eye of the beholder and such tools are sparingly used by economic theorists. Furthermore, computers have already been used to generate numerical examples, providing only approximate, rather than exact, results. This gap opens a window of opportunity for the development of new educational tools directed to socio-economic science students. In a previous work [26], it was shown how some dynamic and interactive mathematical tools, created with *Mathematica*, can be used to promote better student activity and engagement in the learning process. Another work [27] discusses some teaching possibilities offered by the F-Tool concept that can provide an active learning environment in socio-economic science subjects.

The current paper intends to present a new interactive and dynamic mathematical tool for the study of the price elasticity of supply concept, the new PES(Linear)-Tool (see Supplementary Materials), which allows students to change a function's parameter values and get the analytical and graphical results in real time. Furthermore, the interactive and dynamic features of this tool make it suitable to promote an active learning environment and it is available, at no cost, in the Computable Document Format. This format allows the use of the PES(Linear)-Tool, even without an active Wolfram *Mathematica* license (additional information about how to work with the CDF format can be found at http://www.wolfram.com/cdf-player/). The potentialities of the PES(Linear)-Tool will be exhaustively explored to introduce and deal with multiple features of the price elasticity of supply, a central concept in economics. In our opinion, its use in classrooms can promote better student activity and engagement in the learning process, among students enrolled in socio-economic programs.

This paper is structured as follows. After this brief introduction section, Section 2 introduces some basic economic concepts which frame the application of the new tool. Section 3 details the F-tool concept and its application. Section 4 presents the design of the PES(Linear)-Tool. Section 5 is dedicated to some final remarks.

2. Basic Economic Concepts

This section introduces some basic concepts related to the price elasticity of supply.

2.1. The Market Supply Curve and the Market Supply Function of a Good

The producers in a given industry will supply a certain quantity of a produced good at a given price. At this price, the sum of all units gives the total market supply of that good. This corresponds to a point on a curve for the commodity. Continuously changing the price and summing individual supply across all suppliers, we can trace out the market supply curve for the good. That is, a market supply curve of a good shows the total units of that good that are supplied at different prices. More specifically, the short-run market supply curve is the horizontal summation of the individual producers' supply curves, that is:

$$Q(P) = \sum_{i=1}^{n} q_i(P), \tag{1}$$

where n represents the total number of producers in the industry and $q_i(P)$ represents the producer i's supply function.

The Linear Case

Considering a linear specification, the market supply function can be written in the general form

$$Q(P) = \alpha P + \beta, \tag{2}$$

with $\alpha, \beta \in \mathbb{R}$, $\alpha \geqslant 0$ and $P \geqslant \max\left(-\frac{\beta}{\alpha}, 0\right)$. These restrictions are according to the economic theory.

In this paper, we consider the market supply inverse function (when $\alpha > 0$), which can be expressed as

$$P(Q) = aQ + b, \tag{3}$$

with $a = \frac{1}{\alpha}$ and $b = -\frac{\beta}{\alpha}$.

According to the above restrictions, $Q \geqslant \max(\beta, 0)$, that is,

$$Q \geqslant \max\left(-\frac{b}{a}, 0\right). \tag{4}$$

2.2. Measurement and Interpretation of Price Elasticity of Supply

The price elasticity of supply (PES) is a measure used in economics to show the responsiveness of the quantity supplied of a good or service to a change in its price. The elasticity, in a numerical form, is defined as the percentage change in the quantity supplied divided by the percentage change in price, that is,

$$PES(Q) = \lim_{\Delta P \to 0} \frac{Percentage\ change\ in\ quantity\ supplied}{Percentage\ change\ in\ price}. \tag{5}$$

Given that we consider the linear case with $\alpha > 0$ and $a = 1/\alpha$, algebraically, the price elasticity of supply is given by the following expression:

$$PES(Q_0) = \lim_{\Delta P \to 0} \frac{\frac{Q_n - Q_0}{Q_0}}{\frac{P_n - P_0}{P_0}} = \lim_{\Delta P \to 0} \frac{\frac{\Delta Q}{Q_0}}{\frac{\Delta P}{P_0}} = \frac{1}{\lim_{\Delta Q \to 0} \frac{\Delta P}{\Delta Q}} \frac{P_0}{Q_0}, \tag{6}$$

where Q_0 is the (positive) quantity supplied and P represents the price.

So, the expression for the price elasticity of supply can be expressed through the derivative of the function defined by (3) as

$$PES(Q_0) = \frac{1}{\frac{dP}{dQ}(Q_0)} \frac{P_0}{Q_0} = \frac{1}{P'(Q_0)} \frac{P_0}{Q_0}. \tag{7}$$

Obviously, the price elasticity of supply takes only non-negative values. Relatively large values of the PES imply that market supply is responsive to price changes, whereas low values indicate that the supply is not very reactive to price changes.

The elasticity takes the value of zero if the quantity does not react to price changes. In this case, the supply is said to be perfectly inelastic. The elasticity takes a value between 0 and 1 if a price change causes a lower change in the quantity supplied. In this case, the supply is said to be inelastic or rigid with respect to the price. The elasticity takes the value of 1 if a price change causes identical change in the quantity supplied. In this case, the supply is said to have a unitary elasticity. Finally, the elasticity takes a value above 1 if a price change causes a higher change in the quantity supplied. In this case, the supply is said to be elastic with respect to price. The limit case occurs when the elasticity is infinite. In this case, the supply is said to be perfectly elastic.

The Linear Case

Considering a linear specification of the market supply function (2) we get the following expression:

$$PES(Q_0) = \frac{1}{a} \frac{P_0}{Q_0}, \tag{8}$$

that is,

$$PES(Q_0) = 1 + \frac{b}{a} Q_0^{-1}, \tag{9}$$

and the following situations must be considered in the design of a dynamic and interactive tool.

Perfectly elastic supply: The limit case occurs when $a = 0$ and $b > 0$. This corresponds to an infinite PES (see Figure 7).

Remark: In this case the market supply function (2) is not defined since the function (3) is not an invertible function.

Elastic supply: This case occurs when $a > 0$ and $b > 0$. This corresponds to a PES above 1 (see Figure 8).

Unit elastic supply: This case occurs when $a > 0$ and $b = 0$. This corresponds to a PES equal to 1 (see Figures 10–12).

Inelastic supply: This case occurs when $a > 0$ and $b < 0$. This corresponds to a PES below 1 (see Figure 9).

Perfectly inelastic supply: This limit case occurs when $\frac{b}{a} Q_0^{-1} = -1$. This corresponds to a PES equal zero (see Figure 14).

Remark: In this case $\alpha = 0$ in the market supply function (2). So, (2) is not an invertible function.

3. Dynamic and Interactive Tools

Faculty members who regularly use strategies to promote active learning typically find several ways to ensure that students learn the assigned content: promoting the dialog and reflection, promoting the acquisition of new knowledge and the transmission of the acquired knowledge, and doing short-assessments every week.

Currently, several software applications can be (free of charge or for a cost) downloaded from the World Wide Web. In particular, there are many dynamic and interactive tools dealing with some specific economic concepts implemented with the computer algebra system *Mathematica*, which is already available in the Wolfram Demonstrations Project website. In this project (http://demonstrations. wolfram.com) the creators of *Mathematica* promote and divulge globally the innovations designed by its users. Some of these applications provide only analytical information (the *Inflation-Adjusted Yield* tool, available at http://demonstrations.wolfram.com/InflationAdjustedYield/, illustrates how one's investment life planning turns on the net of nominal investment yield and inflation, according to its author). Several other tools provide only graphical information (the *Short-Run Cost Curves* tool, available at http://demonstrations.wolfram.com/ShortRunCostCurves/, provides graphical information about the cubic cost function and its average and marginal cost curves; the *Monopoly Profit and Loss* tool, available at http://demonstrations.wolfram.com/MonopolyProfitAndLoss/, provides graphical information about the marginal cost and the average cost curves). In particular, for the elasticity of demand concept there are tools that provide non-rigorous analytical information such as *The Price Elasticity of Demand* tool (available at http://demonstrations.wolfram.com/ ThePriceElasticityOfDemand/) which shows two ways to calculate the price elasticity of demand), and tools that provide only graphical information (the *Constant Price Elasticity of Demand* tool, available at http://demonstrations.wolfram.com/ThePriceElasticityOfDemand/ illustrates the price elasticity of demand for a specific inverse demand function). However, none of these applications provide all rigorous and exhaustive required information for a global and deep understanding of economic concepts introduced at undergraduate levels, in higher education institutions. Furthermore, these existing materials can hardly be adapted to explain specific concepts in socio-economic sciences or they would require additional resources from both the teacher and the students. This is a gap that the new educational tool described in this paper intends to fulfill since it is adapted to specific training programs to meet educational goals. It allows the design of tasks for independent work and the analysis of individual special cases that are important to recent graduate economists.

3.1. The F-Tool Concept

The F-Tool concept, which was first presented in the 1st National Conference on Symbolic Computation in Education and Research (Portugal 2012), where it was distinguished with the *Timberlake Award for Best Article by a Young Researcher*, was created as an interactive *Mathematica* notebook, specifically to explore the concept of real functions and their graphics, by analyzing the effects caused by changing the values of the parameters of general analytical expressions [28]. Each F-Tool allows the study of a typical class of functions. For each class, a set of parameters is considered such that the class is fully determined by the corresponding analytical expression. This means that each F-Tool provides graphical and rigorous analytical information for all the functions within the corresponding class. In fact, unlike the other tools available in the *Wolfram Demonstrations Project* website, all the tools created under the F-Tool concept provide all the graphical and analytical information desired by the user. Additionally, the user can get exact or approximate analytical results. Finally, the new PES(Linear)-Tool has a very intuitive interface that allows even the most inexperienced user, with no previous knowledge in educational software, to start using all its features in an efficient and autonomous way.

The existing F-Tool are available, free of charge, in the Computable Document Format and the corresponding CDF files can be downloaded for free at https://sapientia.ualg.pt. This format allows anyone with a computer to fully use it, even without an active Wolfram *Mathematica* license.

The F-Tool's framework is composed by three main panels (see Figure 1):

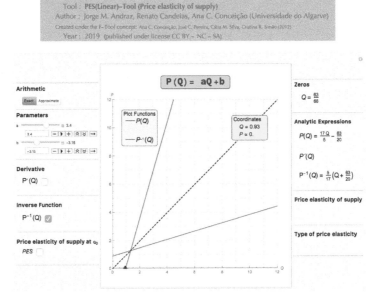

Figure 1. A general example of the price elasticity of supply (PES)(Linear)-Tool: How to get the market supply function in terms of the variable Q.

In the left panel, the user can set the parameters' values, and choose which functions related with the main function are to be displayed.

In the middle panel, all the functions are plotted, according to the options defined in the left panel.

In the right panel, all the analytical information is displayed in accordance with the options chosen by the user in the left panel.

In summary, all the controls and options for all functionalities are located in the left panel. As the user interacts dynamically with the tool, all the graphical and analytical results are displayed in real time in the middle and right panels, respectively. When choosing the option ►, the user will then see the corresponding graphics moving continuously and the analytical information changing accordingly. It is through this kind of dynamic interaction that "computer algebra systems present new opportunities for teaching and learning" [29].

The use of the F-Tool concept in the classroom allows a dynamic approach to various concepts related to the study of functions and promotes new ways of reasoning/thinking, evaluating, teaching, and learning. The F-Tool concept was conceived as an active learning tool, that is, its adequate use provides a context of teaching and learning where students and teachers are both invited to fully participate [30]. Through dynamic changes of the parameters values, it is possible to obtain rigorous analytical information, presented in exact or approximate arithmetic, as well as static and non-static visual information [22]. Although it is a dynamic and interactive educational software, the F-Tool can also be used in the construction of multiple choice and open response evaluation questions [1].

3.2. The F-Tool Concept Adapted to the Socio-economic Sciences

Taking into account our experience of using dynamic and interactive mathematical tools [1] as active learning tools in natural science courses, we decided to adapt this type of approach to some economic concepts. The idea is to focus the teaching process on the students, stimulating their participation and motivating those with a level of math knowledge, often insufficient, to obtain new knowledge in a solid way. In this way, it becomes possible to teach new concepts in a solid and consistent way.

The most common way for faculty members to engage students in active learning is by stimulating the discussion [20]. A variety of materials and techniques can be used to trigger the discussion and each teacher can provide several experiences that will stimulate the discussion among students. Demonstrations during a lecture can be used to stimulate the students' curiosity and to improve their understanding of conceptual material and processes [31], particularly when the demonstration invites students to participate in research activities through the use of questions such as "What would happen if we change dynamically the parameter b? Would the price elasticity of supply change? And what would happen if the parameter a changes dynamically?" (see Figures 8–10). So, the faculty member can encourage the discussion, dialogue, and reflection in the classroom, proposing stimulating exercises that lead to a supervised constructive debate among the students.

In Section 4 we present the new dynamic and interactive economic tool, called the PES(Linear)-Tool, created under the F-Tool concept. The usefulness of this tool is illustrated by introducing the price elasticity of supply concept in a microeconomics class, as well as all the analytical and graphical information involved with the analysis of this concept.

4. Designing the New PES(Linear)-Tool

The use of the symbolic computation capabilities of *Mathematica*, and its own programming language (along with the pretty-print functionality that allows one to write mathematical expressions on the computer using the traditional notation, as on paper), enables us to implement on a computer, and in a rather straightforward manner, all the ideas that go into the F-Tool concept.

The PES(Linear)-Tool was created as an interactive *Mathematica* notebook and it is available online, in the Computable Document Format, as a supplement to this article. It allows the exploration of concepts related to a market supply function (3), where $a, b \in \mathbb{R}$, $a > 0$ and $Q > \max\left(-\frac{b}{a}, 0\right)$. It should be noted that the particular case of $a = 0$ was also included to exemplify the perfectly elastic supply (when $Q > 0$) (see Figure 7). In terms of implementation and in spite of their mathematic simplicity, constant functions should be dealt with separately because they have no inverse function (see Figure 7). This means that the constant case has to be coded separately, in order to generate the correct analytical information for those functions. The PES(Linear)-Tool provides all graphical and analytical information of the inverse function of $P(Q)$ (that is, the market supply function). As students often confuse the concepts of elasticity and derivative, the tool provides the option "Derivative" on the left panel (see Figures 10–12). The PES(Linear)-Tool displays graphical information on the value of the $PES(Q_0)$ whenever this option is selected. This allows the user to visualize the change from an economic model with an elastic supply to a model with an inelastic supply (going through a unitary elastic supply). As in the F-Tool, the user can interact with this information in real time.

As an illustration of this tool, let us to consider the plot of the inverse function (3) as depicted in Figure 1, and the market supply function (in terms of the variable *Q*). In this case, the exact analytical expressions of the function and its inverse are displayed, once the exact arithmetic option has been selected. The dashed line displayed on the plot is described by the equation $y = x$ and corresponds to the symmetry axis of the inverse transformation.

The PES(Linear)-Tool is essentially created by a single `Manipulate` command (see Figure 2), whose output is not just a static result but a running program that we can interact with. In fact, the code consists of some initial definitions followed by the single command `Manipulate`. This command is responsible for creating the interactive object that contains the three panels. In particular, the command `Manipulate` generates all the functional controls, such as the sliders for the parameters' values and checkboxes for the plots' options. Through dynamic changes of the parameters' values, it is possible to obtain approximate or exact analytical information, as well as static and non-static visual information [28].

```
Manipulate[

  Labeled[

    DynamicModule[[ ...] PlotStyle → {{ColorData[54][17], Thick}, [ ...][ ...][ ...][ ...]]
```

Figure 2. General code structure of the PES(Linear)-Tool.

To create the PES(Linear)-Tool we used part of the code of the educational software F-Tool. Obviously, to provide all the graphical and analytical information for the price elasticity of supply, several adaptations were performed and new fields related to this socio-economic concept were added. Figure 3 displays the code block that generates the value of the price elasticity of supply at a given quantity Q_0. It should be noticed that the cases of $a \leqslant 0$ and/or $Q_0 \leqslant \max\left(-\frac{b}{a}, 0\right)$ (see Figures 5, 6, 13) should be considered separately.

```
    Style["Price elasticity of supply at Q₀", Bold],

  Row[{"      ", Control[{{elasticityD, False, "PES"}, {True, False}}]}],

  [ ...]
      Dynamic[TradForm[If[elasticityD,

    If[a < 0, Row[{"    Does not make sense", "\n    to analyze PES(Q₀) !"}],

      [ ...]
      If[( a > 0) && ((-b / a) > 0) && (Q0 <= -b / a), Row[{""}],

      [ ...]
        If[RPES[Q0, a, b] > 1, Row[{"    Elastic Supply"}]
```

Figure 3. Code snippet of the PES(Linear)-Tool. This is part of the code that generates the analytical information about the price elasticity of supply.

4.1. Parameters a and b

In order to create a consistent tool that considers all the mathematical possibilities for which the economic model makes sense, several situations concerning the values of the parameters *a* and *b* should be implemented. Given the function (3), only non-negative values for the parameter *a* are considered in the code (the range of values that run through the slider, see Figure 4).

```
Style["Parameters", Bold], "",

{{a, 1}, 0, 5, 0.01, ImageSize → Small, Appearance → "Labeled"},

{{b, 0}, -12, 12, 0.01, ImageSize → Small, Appearance → "Labeled"}
```

Figure 4. Code snippet of the PES(Linear)-Tool. This is part of the code that generates the range of values for the parameter *a* that run through the slider.

The user can also introduce directly the parameters' values. However, for certain values of the parameters, the correspondent market supply function is not defined and therefore, the PES(Linear)-Tool will exhibit the following message: "Does not make sense to analyze PES(Q0)!", whenever the PES button is selected (see Figure 5). Consequently, all options will be unavailable until acceptable parameter values are considered. This situation occurs when the user chooses a non-positive value for the parameter *a*, and/or the user chooses a non-positive value for the variable *Q* (see Figures 6 and 13).

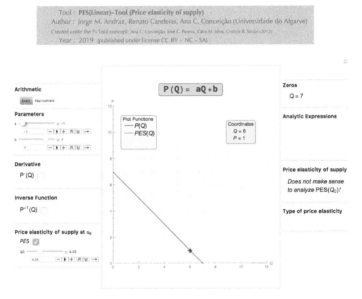

Figure 5. An example of the information obtained when a negative value for the parameter *a* is chosen.

Depending on the values considered for the parameters *a* and/or *b*, *Q* can assume values in different numeric sets. So, the values of Q_0 that can be considered depend on the values of *a* and *b*. The PES(Linear)-Tool can be used to improve the students' understanding of this conceptualization because it enables students to analyze the relationship between the null value of the function $P(Q)$ and the range of acceptable values for Q_0 (see Figure 6).

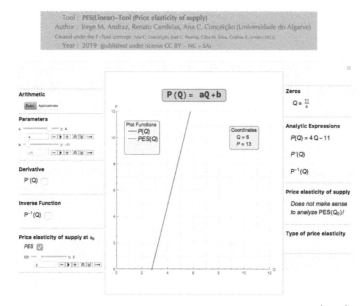

Figure 6. Example of a non economic model due the fact that $Q_0 \leqslant \max\left(-\frac{b}{a}, 0\right)$.

4.2. Perfectly Elastic Supply

This subsection illustrates how the PES(Linear)-Tool can be used to improve the students' understanding of the price elasticity of supply concept (see Figure 7). In the classroom the faculty can explain that this is a limit case that occurs when the market supply function is not defined. The teacher may ask the students if a change in the parameter b causes any change in the type of price elasticity. Then it can be asked about the effects of a possible change in parameter a.

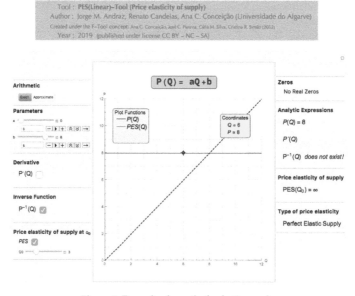

Figure 7. Example of a perfectly elastic supply.

4.3. Elastic and Inelastic Supplies

This subsection describes how the PES(Linear)-Tool can be used to improve the students' understanding of the price elasticity of supply concept.

By using the PES(Linear)-Tool the faculty can ask the students to interpret the value $PES(Q_0)$ depending on the values of a, b, and Q_0. The faculty can start with an example of an elastic supply and ask the students to identify the parameter to be changed in order to get an inelastic supply and how the value of Q_0 affects the elasticity's value (see Figures 8–10).

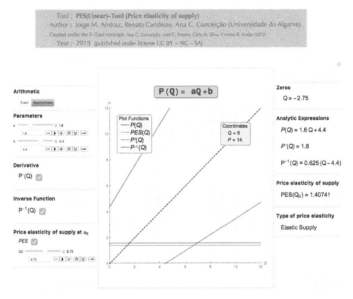

Figure 8. Example of an elastic supply.

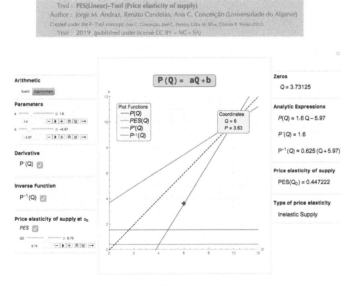

Figure 9. Example of an inelastic supply.

4.4. Unit Elastic Supply

It is generally acknowledged that there is often a confusion between the concepts of elasticity and derivative among students. In order to illustrate the contribution of the PES(Linear)-Tool to distinguish such concepts, this subsection presents some examples of unit elastic supply functions associated to different derivatives' values. Figures 10–12 present examples of unitary elasticity supply functions associated to derivatives' values above, below and equal to 1, respectively.

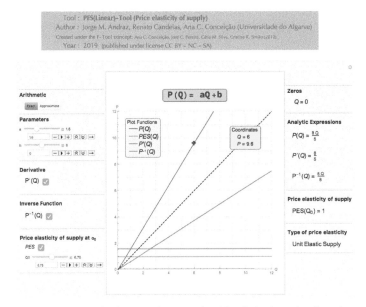

Figure 10. Example of an unit elastic supply with a derivative value above 1.

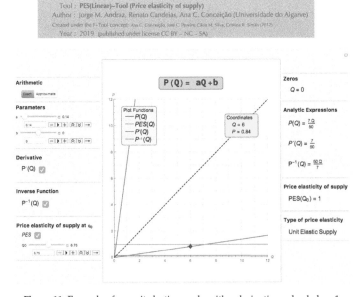

Figure 11. Example of an unit elastic supply with a derivative value below 1.

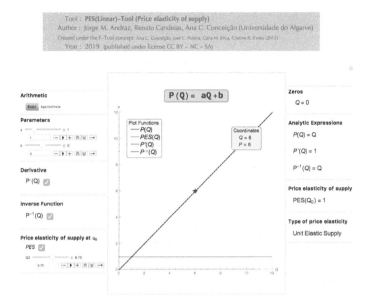

Figure 12. Example of an unit elastic supply with a derivative value equal to 1.

The graphical and analytical information, reported by the tool, confirm that despite the existence of a relationship between the two concepts (elasticity and derivative), their values are not directly connected.

Finally, Figure 13 exhibits a non economic model in which $Q_0 \leqslant \max\left(-\frac{b}{a}, 0\right)$ (if any positive value of Q_0 is considered, the economic model would have a unitary elasticity).

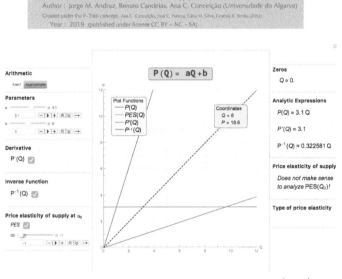

Figure 13. Example of a non economic model due the fact that $Q_0 \leqslant \max\left(-\frac{b}{a}, 0\right)$ (if any positive value of Q_0 is considered the economic model would have a unitary elasticity).

4.5. Perfectly Inelastic Supply

Although the PES(Linear)-Tool cannot fully illustrate the perfectly inelastic supply case, it can be used to make this case easier for students to understand. Once the information that this limit case occurs when $\frac{b}{a}Q_0^{-1} = -1$ has been transmitted to the students, the immediate conclusion is that the corresponding PES is zero. In this case, the teacher should state that α, in (2), is also null (or question why) and therefore (2) is not an invertible function. This case is depicted in Figure 14.

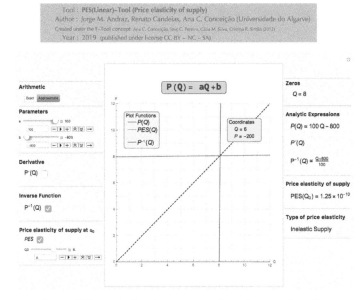

Figure 14. Example of an almost perfectly inelastic economic model.

5. Final Remarks

This paper presents a new dynamic and interactive tool created with the computer algebra system *Mathematica*, the PES(Linear)-Tool, designed to be applied in economics education, a domain where the use of computer algebra has been particularly limited. Although there are several free of charge applications available at the Wolfram Demonstrations Project website, none of those provide all the graphical and analytical information necessary for a good understanding of the concepts introduced in socio-economic undergraduate courses in universities. Moreover, these applications provide either graphical or analytical information, but not both, and/or only for some particular cases, and they can hardly be adapted to explain specific concepts in social economic sciences, or that adaptation would require additional resources from both the teacher and the students.

The above mentioned issues constitute several gaps which the new tool intends to fulfill. The PES(Linear)-Tool is a computer algebra tool directed to the study of one of the most widely used concepts in socio-economics courses—the price elasticity of supply. Starting with the specification of the market supply function, the design, functionalities and capabilities of the PES(Linear)-Tool are exhaustively explored in this paper to analyze the price elasticity of supply, accounting for all the mathematical possibilities for which the economic model makes sense. This tool also differs from other existing tools in that it can be downloaded at no cost and allows the complete analysis of multiple situations involving the study of the price elasticity supply in a dynamic and interactive way. Specifically, the tool offers the students the possibility of changing the parameters' values in the economic model and getting both the analytical and graphical effects in real time. The new PES(Linear)-Tool has a very intuitive interface that allows even the most inexperienced user, with no previous knowledge in educational software, to start using all the features in an efficient and autonomous way.

Given the recognition in the literature that some economic concepts can be more easily understood when students work with a large number and variety of graphics in an interactive way, with the support of the appropriate technology, we believe that the use of the PES(Linear)-Tool in the classroom can promote new ways of reasoning/thinking, evaluating, teaching, and learning in a context where students and teachers are invited to contribute. In this way, this tool promotes the active learning in classrooms and simultaneously students' autonomous work, by allowing the design of challenging problems based on dynamic and interactive exercises using the CDF format, which students can work on and then send in their results to the faculty by email.

The design of the PES(Linear)-Tool can be generalized to other economic models that can be studied through other classes of functions, and also opens the possibility for the development of other interactive tools associated with other economic concepts.

Going forward, a statistically rigorous study in loco to assess the students' perception when using the PES(Linear)-Tool, and therefore to estimate the tool's pedagogical value is of extremely importance. We believe that this study can be an important help for the future development of these kind of educational tools.

Supplementary Materials: The PES(Linear)-Tool is available online at http://www.mdpi.com/2297-8747/24/4/87/s1.

Author Contributions: The dynamic and interactive tool presented in this paper was designed by A.C.C. (under the F-Tool concept created by A.C.C., José C. Pereira, Cátia M. Silva, and Cristina R. Simão). All authors contributed to the improvement of the tool. The implementation with the computer algebra system *Mathematica* was made by R.C. and A.C.C. The conceptualization and methodology was performed by J.M.A. and A.C.C. The paper was written by J.M.A. and A.C.C. All authors reviewed the manuscript.

Funding: This research was funded by Fundação para a Ciência e a Tecnologia within the project UID/ECO/04007/2019.

Acknowledgments: The authors thank the contributions and suggestions of two anonymous referees.

Conflicts of Interest: The authors declare no conflict of interest.

References

1. Conceição, A.C.; Coelho, A.C.; Gonçalves, C.D. Estratégia pedagógica no Ensino Superior baseada no conceito de aprendizagem ativa. In *Ensino-Aprendizagem de Ciências e Suas Tecnologias*; Schimiguel, J., Frenedozo, R.C., Coelho, A.C., Eds.; Edições Brasil: Jundiaí, Brasil, 2019; pp. 9–25.
2. Conceição, A.C.; Pereira, J.C.; Silva, C.M.; Simão, C.R. Software educacional em pré-cálculo e cálculo diferencial: O conceito F-Tool. In Proceedings of the Encontro Nacional da SPM 2012, Faro, Portugal, 9–11 July 2012; pp. 57–60.
3. Costanzo, F.; Gray, G. On the implementation of interactive dynamics. *Int. J. Eng.* **2000**, *16*, 385–393.
4. Foertsch, J.; Moses, G.; Strikwerda, J.; Litskow, M. Reversing the lecture/homework paradigm using eTEACH web-based streaming video software. *Int. J. Eng.* **2002**, *91*, 267–274. [CrossRef]
5. Fogler, H.S.; Montomery, S.M.; Zipp, R.P. Interactive computer modules for undergraduate chemical engineering instruction. *Comput. Appl. Eng. Educ.* **1996**, *1*, 11–24. [CrossRef]
6. García, O.; Laredo, M. Comunidades Académicas Virtuales Como Medio en la Enseñanza y Aprendizaje Usando Software Matemático. 2014. Available online: http://www.pag.org.mx/index.php/PAG/article/view/93 (accessed on 8 October 2019).
7. Gray, G.; Costanzo, F. The interactive classroom and its integration into the mechanics curriculum. *Int. J. Eng.* **1999**, *15*, 41–50.
8. Guamán, L.R.B.; Córdova, C.C. Using Wolfram software to improve reading comprehension in mathematics. In Proceedings of the 2016 EBMEI International Conference on Education, Information and Management (EBMEI-EIM 2016), São Paulo, Brazil, 31 August–1 September 2016; pp. 53–58.

9. Morales, F.; Valencia, A.; Valencia, R.; Mario, J. Análisis de software matemático usados en nivel superior. *Rev. Vínculos* **2013**, *10* , 299–307.

10. Prado, J.L.; Freira, A.M.; Albuquerque, I.; Júior, P.P. Experienciando o software *Mathematica* na sala de aula. In Proceedings of the IV Colóquio Internacional Educação e Contemporaneidade, Laranjeiras, Brasil, 22–24 September 2010.

11. Randow, C.L.; Miller, A.J.; Costanzo, F.; Gray, G.L. *Mathematica* Notebooks for Classroom Use in Undergraduate Dynamics: Demonstration of Theory and Examples. In Proceedings of the 2003 American Society for Engineering Education Annual Conference & Exposition, Nashville, TN, USA, 22–25 June 2003.

12. Silva, J.; Astudillo, A. CbL-Cálculo: Curso b-Learning Para el Apoyo de la Enseñanza y Aprendizaje de Cálculo en Ingeniería. 2012. Available online: http://revistas.um.es/red/article/view/232581 (accessed on 8 October 2019).

13. Macintyre, T.; Forbes, I. Algebraic skills and CAS—Could assessment sabotage the potential? *Int. J. Comput. Algebra Math. Educ.* **2002**, *9*, 29–56.

14. Buchberger, B. Should students learn integration rules? *ACM SIGSAM Bull.* **1990**, *24*, 10–17. [CrossRef]

15. Kilicman, A.; Hassan, M.A.; Said Hussain, S.K. Teaching and learning using mathematics software "The New Challenge". *Procedia Soc. Behav. Sci.* **2010**, *8*, 613–619. [CrossRef]

16. Mason, J. A Comprehensive Mathematics curriculum with *Mathematica*. Available online: https://library.wolfram.com/infocenter/Conferences/5360/ (accessed on 8 October 2019).

17. Ramos, A.; Delgado, F.; Afonso, P.; Cruchinho, A.; Pereira, P.; Sapeta, P.; Ramos, G. Implementação de novas práticas pedagógicas no Ensino Superior. *Revista Portuguesa de Educação* **2013**, *26*, 115–141. [CrossRef]

18. Meyers, C.; Jones, T.B. Promoting active learning: Strategies for the college classroom. *Biochem. Educ.* **1994**, *2*, 192.

19. Chan, M.M.; Amado-Salvatierra, H.R.; Plata, R.B.; Hernández Rizzardini, R. La efectividad del uso de simuladores para la construcción de conocimiento en un contexto MOOC. In Proceedings of the II International Conference MOOC-Maker (MOOC-Maker 2018), Medellín, Colombia, 11–12 October 2018; pp. 42–53.

20. Bonwel, C.C.; Eison, J.A. Active Learning: Creating excitement in the classroom. Available online: https://eric.ed.gov/?id=ED336049 (accessed on 8 October 2019).

21. Cross, K.P.; Angelo, T.A. *Classroom Assessment Techniques: A Handbook for Faculty*; National Center for Research to Improve Postsecondary Teaching and Learning: Ann Arbor, MI, USA, 1988.

22. Conceição, A.C. Software educativo em pré-cálculo e cálculo diferencial. *Rev. Ciênc. Elem.* **2018**, *6*, 36–38. [CrossRef]

23. Mulligan, C.B. Quantifier elimination for deduction in econometrics. Available online: https://www.nber.org/papers/w24601 (accessed on 8 October 2019).

24. Mulligan, C.B. Automated economic reasoning with quantifier elimination. Available online: https://www.nber.org/papers/w22922 (accessed on 8 October 2019).

25. Mulligan, C.B.; Bradford, R.; Davenport, J.H.; England, M.; Tonks, Z. Non-linear real arithmetic benchmarks derived from automated reasoning in economics. Available online: https://www.nber.org/papers/w24602 (accessed on 8 October 2019).

26. Andraz, J.M.; Conceição, A.C. Dynamic and interactive mathematical tools in socio-economics sciences classrooms. In Proceedings of the 4th International Conference on Numerical and Symbolic Computation Developments and Applications (SYMCOMP2019), Porto, Portugal, 11–12 April 2019; pp. 321–336.

27. Andraz, J.M.; Candeias, R.; Conceição, A.C.; Serafim, I. An interactive way of analyzing economic concepts using symbolic computation. In Proceedings of the 4th International Conference on Numerical and Symbolic Computation Developments and Applications (SYMCOMP2019), Porto, Portugal, 11–12 April 2019; pp. 343–356.

28. Conceição, A.C.; Pereira, J.C.; Silva, C.M.; Simão, C.R. *Mathematica* in the classroom: New tools for exploring precalculus and differential calculus. In Proceedings of the 1st National Conference on Symbolic Computation in Education and Research (CSEI 2012), Lisboa, Portugal, 2–3 April 2012.

29. Hayden, M.B.; Lamagna, E.A. NEWTON: An interactive environment for exploring mathematics. *J. Symb. Comput.* **1998**, *25*, 195–212. [CrossRef]

30. Conceição, A.C.; Fernandes, S.; Pereira, J.C. Prática pedagógica com o software educacional F-Tool em Cálculo I. In Proceedings of the Congresso Nacional de Práticas Pedagógicas no Ensino Superior (CNaPPES 2015), Leiria, Portugal, 3 July 2015; pp. 99–104.
31. Shakhashiri, B.Z. Lecture demonstrations. *J. Chem. Educ.* **1984**, *61*, 1010–1011. [CrossRef]

Mathematical and Computational Applications

Article

The Invariant Two-Parameter Function of Algebras $\bar{\psi}$

José María Escobar [1], Juan Núñez-Valdés [1,*] and Pedro Pérez-Fernández [2,3]

[1] Dpto. de Geometría y Topología, Facultad de Matemáticas, Universidad de Sevilla, Calle Tarfia s/n, 41012 Sevilla, Spain; pinchamate@gmail.com

[2] Dpto. de Física Aplicada III, Escuela Técnica Superior de Ingeniería, Universidad de Sevilla, Camino de los Descubrimientos, 41092 Sevilla, Spain; pedropf@us.es

[3] Instituto Carlos I de Física Teórica y Computacional, Universidad de Granada, Fuentenueva s/n, 18071 Granada, Spain

* Correspondence: jnvaldes@us.es

Received: 12 September 2019; Accepted: 11 October 2019; Published: 14 October 2019

Abstract: At present, the research on invariant functions for algebras is very extended since Hrivnák and Novotný defined in 2007 the invariant functions ψ and φ as a tool to study the Inönü–Wigner contractions (IW-contractions), previously introduced by those authors in 1953. In this paper, we introduce a new invariant two-parameter function of algebras, which we call $\bar{\psi}$, as a tool which makes easier the computations and allows researchers to deal with contractions of algebras. Our study of this new function is mainly focused in Malcev algebras of the type Lie, although it can also be used with any other types of algebras. The main goal of the paper is to prove, by means of this function, that the five-dimensional classical-mechanical model built upon certain types of five-dimensional Lie algebras cannot be obtained as a limit process of a quantum-mechanical model based on a fifth Heisenberg algebra. As an example of other applications of the new function obtained, its computation in the case of the Lie algebra induced by the Lorentz group $SO(3,1)$ is shown and some open physical problems related to contractions are also formulated.

Keywords: invariant functions; contractions of algebras; Lie algebras; Malcev algebras; Heisenberg algebras

1. Introduction

Regarding the concept of *limit process* between physical theories in terms of contractions of their associated symmetry groups, formulated by Erdal Inönü and Eugene Wigner [1,2], these authors introduced the so-called *Inönü–Wigner contractions (IW-contractions)* in 1953. Later, other extensions of these IW-contractions have also been addressed, for instance the *generalized Inönü–Wigner contractions*, introduced by Melsheiner [3], the *parametric degenerations* [4–6], widely used in the Algebraic Invariants Theory, and the *singular contractions* [2]. To study these contractions, Hrivnák and Novotný introduced the invariant functions ψ and φ as a tool in 2007 [7]. These invariant functions depend on one parameter.

Continuing with this topic, the main goal of this paper is to introduce a new invariant function, in this case depending on two parameters, which we call *the two-parameter invariant function* $\bar{\psi}$, to get some advances on this research. Indeed, the objective is to prove, by means of this function, that the five-dimensional classical-mechanical model built upon certain types of five-dimensional Lie algebras cannot be obtained as a limit process of a quantum-mechanical model based on a fifth Heisenberg algebra.

Indeed, the study of this function is mainly focused in the frame of the Malcev algebras of the type Lie. Thus, this paper can be considered as the natural continuation of a previous one dealing with Lie algebras [8]. We try to generalize the properties obtained on that to the case of Malcev algebras.

Math. Comput. Appl. **2019**, *24*, 89; doi:10.3390/mca24040089   www.mdpi.com/journal/mca

Math. Comput. Appl. **2019**, *24*, 89

The structure of the paper is as follows. In Section 2, we recall some preliminaries on the mathematical objects dealt with in this paper, Lie algebras and Malcev algebras. Section 3 is devoted to introducing and proving the main properties of the two-parameter invariant function $\bar{\psi}$. For computations, we used the SAGE symbolic computation package and in this section we prove that this new function is different from others previously defined, which are used as a tool to study contractions of algebras. We also prove the main result of the paper: no proper contraction between a fifth Heisenberg algebra and a filiform Lie algebra of dimension 5 exists. It implies that the five-dimensional classical-mechanical model built upon a five-dimensional filiform Lie algebra cannot be obtained as a limit process of a quantum-mechanical model based on a fifth Heisenberg algebra. In this way, the new function allows us to step forward in the research on contractions. In Section 4, we show some of our discussion and conclusions regarding the research done. Finally, in Section 5, we give some comments on the materials and methods used in such a research.

2. Preliminaries

We show in this section some preliminaries on Lie algebras, Malcev algebras and on Heisenberg algebras, which are the main mathematical objects used in the paper.

2.1. Preliminaries on Lie Algebras

In this subsection, we show some preliminaries on Lie algebras. For a further review on this topic, the reader can consult [9].

An n-dimensional *Lie algebra* $\mathfrak{g}$ over a field K is an n-dimensional vector space over K endowed with a second inner law, named *bracket product*, which is bilinear and anti-commutative and satisfies the *Jacobi identity*

$$J(u, v, w) = [u, [v, w]] + [v, [w, u]] + [w, [u, v]] = 0, \text{ for all } u, v, w \in \mathfrak{g}. \tag{1}$$

The law of the n-dimensional Lie algebra $\mathfrak{g}$ is determined by the products

$$[e_i, e_j] = \sum_{k=1}^{n} c_{ij}^k e_k, \quad \text{for} \quad 1 \leq i < j \leq n,$$

where $c_{i,j}^k \in K$ are called *structure constants* of $\mathfrak{g}$. If all these constants are zero, then the Lie algebra is called *abelian*.

Two Lie algebras $\mathfrak{g}$ and $\mathfrak{h}$ are *isomorphic* if there exists a vector space isomorphism f between them such that $f([u, v]) = [f(u), f(v)]$, for all $u, v \in \mathfrak{g}$.

A mapping $d : \mathfrak{g} \longrightarrow \mathfrak{g}$ is a *derivation* of $\mathfrak{g}$ if $d([u, v]) = [d(u), v] + [u, d(v)]$, for all $u, v \in \mathfrak{g}$. The set of derivations of $\mathfrak{g}$ is denoted by $Der\mathfrak{g}$.

The *lower central series* of a Lie algebra $\mathfrak{g}$ is defined as $\mathfrak{g}^1 = \mathfrak{g}$, $\mathfrak{g}^2 = [\mathfrak{g}^1, \mathfrak{g}]$, ..., $\mathfrak{g}^k = [\mathfrak{g}^{k-1}, \mathfrak{g}]$, ...

If there exists $m \in \mathbb{N}$ such that $\mathfrak{g}^m \equiv 0$, then $\mathfrak{g}$ is called *nilpotent*. The *nilpotency class* of $\mathfrak{g}$ is the smallest natural c such that $\mathfrak{g}^{c+1} \equiv 0$.

An n-dimensional nilpotent Lie algebra $\mathfrak{g}$ is said to be *filiform* if it is verified that dim $\mathfrak{g}^k = n - k$, for all $k \in \{2, \ldots, n\}$. Filiform Lie algebras were introduced by Vergne in her Ph.D. Thesis, in 1966, later published in [10] in 1970.

The only n-dimensional filiform Lie algebra for $n < 3$ is the abelian. For $n \geq 3$, it is always possible to find an *adapted basis* $\{e_1, \ldots, e_n\}$ of $\mathfrak{g}$ such that $[e_1, e_2] = 0$, $[e_1, e_j] = e_{j-1}$, for all $j \in \{3, \ldots, n\}$ and $[e_2, e_j] = [e_3, e_j] = 0$, for all $j \in \{3, \ldots, n\}$.

From the condition of filiformity and the Jacobi identity in Equation (1), the bracket product of $\mathfrak{g}$ is determined by

$$[e_i, e_j] = \sum_{k=2}^{min\{i-1,n-2\}} c_{ij}^k e_k, \quad \text{for} \quad 4 \le i < j \le n,$$

where $c_{i,j}^k \in K$ are called *structure constants* of $\mathfrak{g}$. If all these constants are zero, then the filiform Lie algebra $\mathfrak{g}$ is called *model*. The model algebra is not isomorphic to any other algebra of the same dimension and every n-dimensional filiform Lie algebra $\mathfrak{g}$ having an adapted basis $\{e_1, \ldots, e_n\}$ verifies that $\mathfrak{g}^2 = \langle e_2, \ldots, e_{n-1} \rangle$, $\mathfrak{g}^3 = \langle e_2, \ldots, e_{n-2} \rangle, \ldots, \mathfrak{g}^{n-1} = \langle e_2 \rangle$, $\mathfrak{g}^n = 0$.

2.2. Preliminaries on Malcev Algebras

Now, we recall some preliminary concepts on Malcev algebras, taking into account that a general overview can be consulted in [11]. From here on, we only consider finite-dimensional Malcev algebras over the complex number field $\mathbb{C}$.

A *Malcev algebra* $\mathcal{M}$ is a vector space with a second bilinear inner composition law $([\cdot, \cdot])$ called the *bracket product* or *commutator*, which satisfies: (a) $[u, v] = -[v, u]$, $\forall u, v \in \mathcal{M}$; and (b) $[[u, v], [u, w]] = [[[u, v], w], u] + [[[v, w], u], u] + [[[w, u], u], v]$, $\forall u, v, w \in \mathcal{M}$. Condition (b) is named *Malcev identity* and we use the notation $M(u, v, w) = [[u, v], [u, w]] - [[[u, v], w], u] - [[[v, w], u], u] - [[[w, u], u], v]$.

Given a basis $\{e_i\}_{i=1}^n$ of a n-dimensional Malcev algebra $\mathcal{M}$, the *structure constants* $c_{i,j}^h$ are defined as $[e_i, e_j] = \sum_{h=1}^n c_{i,j}^h e_h$, for $1 \le i, j \le n$.

It is immediate to see that Malcev algebras and Lie algebras are not disjoint sets. Indeed, every Lie algebra is a Malcev algebra, but the converse is not true. Therefore, we can distinguish between Malcev algebras of the type Lie and Malcev algebras of the type non-Lie. Obviously, those Malcev algebras which are of the type Lie verify both identities: Jacobi and Malcev.

If the Jacobi identity does not hold, then the Malcev algebra is said to have a *Jacobi anomaly*. In quantum mechanics, the existence of Jacobi anomalies in the underlying non-associative algebraic structure related to the coordinates and momenta of a quantum non-Hamiltonian dissipative system was already claimed by Dirac [12] in the process of taking Poisson brackets. In string theory, for instance, one such anomaly is involved by the non-associative algebraic structure that is defined by coordinates $(\vec{x})$ and velocities or momenta $(\vec{v})$ of an electron moving in the field of a constant magnetic charge distribution, at the position of the location of the magnetic monopole [13]. In particular, $J(v_1, v_2, v_3) = -\vec{\nabla} \circ \vec{B}(\vec{x})$, where $\vec{\nabla} \circ \vec{B}(\vec{x})$ denotes the divergence of the magnetic field $\vec{B}(\vec{x})$. The underlying algebraic structure constitutes a non-Lie Malcev algebra [14], with the commutation relations $[x_a, x_b] = 0$, $[x_a, v_b] = i \, \delta_{ab}$ and $[v_a, v_b] = i \, \varepsilon_{abc} \, B_c(\vec{x})$, where $a, b, c \in \{1, 2, 3\}$, δ_{ab} denotes the Kronecker delta and ε_{abc} denotes the Levi–Civita symbol. If the magnetic field is proportional to the coordinates, the latter can be normalized and $B_c(\vec{x})$ can then be supposed to coincide with x_c. The resulting algebra is then called magnetic [15]. A generalization to electric charges has recently been considered [15] by defining the products $[x_a, x_b] = -i \, \varepsilon_{abc} \, \vec{E}_c(\vec{x}, \vec{v})$, where the electric field $\vec{E}$ as well as the magnetic field $\vec{B}$ must depend not only on coordinates but also on velocities. It is worth remarking that both magnetic and electric algebras constitute magma algebras (see [16] for this last concept).

If $\mathfrak{g}$ is a Malcev algebra of the type Lie and $D \in Der\mathfrak{g}$ a derivation of $\mathfrak{g}$, then, according to the anti-commutative property of $\mathfrak{g}$ and the Jacobi identity in Equation (1) of Lie algebras, we get that

$$[d[x, y], [x, z]] + [[x, y], d[x, z]] = d[[[x, z], y], x] + d[[[z, x], x], y] \quad \forall x, y, z \in \mathfrak{g}$$

Starting from here and due to reasons of length, only Malcev algebras of type Lie, that is to say, actually Lie algebras, are used in this paper. Malcev algebras of type non-Lie will be dealt with in future work.

2.3. Preliminaries on Heisenberg Algebras

Let n be a non-negative integer or infinity. The nth Heisenberg algebra (so-called after Werner Karl Heisenberg) is the Lie algebra with basis $\mathcal{B} = \{p_1, \ldots, p_n, q_1, \ldots, q_n, z\}$ with the following relations, known as *canonical commutation relations*

1. $[p_i, q_j] = c_{ij} z, \quad 1 \leq i, j \leq n.$
2. $[p_i, z] = [q_i, z] = [p_i, p_j] = [q_i, q_j] = 0, \quad 1 \leq i, j \leq n.$

Note that the dimension of an nth Heisenberg algebra is not n, but $2n + 1$. In fact, the n in the above definition is called the *rank* of the Heisenberg algebra, although it is not, however, a rank in any of the usual meanings that this word has in the theory of Lie algebras. Thus, this Lie algebra is also known as the *Heisenberg algebra of rank n*.

In any case, from here on and to avoid confusions we designate under the notation fifth Heisenberg algebras to those Heisenberg algebras generated by five generators.

3. Results

In this section, which is divided by subheadings, we provide a concise and precise description of our experimental results. They are the following.

3.1. Introducing a New Invariant Function

Let $\mathfrak{g} = (V, [\,,])$ be a Lie algebra. *End* $\mathfrak{g}$ denotes the vector space of all linear operators of $\mathfrak{g}$ over V.

Definition 1. *Let $\mathfrak{g}$ be a Lie algebra. The set*

$$Der_{(\alpha,\beta,\gamma,\tau)}\mathfrak{g} = \{d \in End\,\mathfrak{g} : \alpha[d[x,y],[x,z]] + \beta[[x,y],d[x,z]] = \gamma d[[[x,z],y]x] + \tau d[[[z,x],x],y]\}$$

$\forall(\alpha, \beta, \gamma, \tau) \in \mathbb{C}^4$, *is called* the set of the $(\alpha, \beta, \gamma, \tau)$-derivations *of the algebra $\mathfrak{g}$. It is denoted by $Der_{(\alpha,\beta,\gamma,\tau)}\mathfrak{g}$.*

It is obvious that $\dim(Der_{(1,1,1,1)}\mathfrak{g}) = \dim(Der\mathfrak{g})$. Then, as $\dim(Der\mathfrak{g})$ is an invariant of $\mathfrak{g}$, it follows that $\dim(Der_{(1,1,1,1)}\mathfrak{g})$ is an invariant of $\mathfrak{g}$. This leads the following result.

Proposition 1. *If $\mathfrak{g}$ is a Lie algebra, then $\dim_{(1,1,1,1)}\mathfrak{g}$ is an algebraic invariant of $\mathfrak{g}$.* □

Theorem 1. *Let $\mathfrak{g}$ and $\bar{\mathfrak{g}}$ be two Malcev algebras of the type Lie and let $f : \mathfrak{g} \to \bar{\mathfrak{g}}$ be an isomorphism. Then, the mapping $\rho : End\,\mathfrak{g} \to End\,\bar{\mathfrak{g}}$, defined by $D \longrightarrow fDf^{-1}$, is an isomorphism between the vector spaces $Der_{(\alpha,\beta,\gamma,\tau)}\mathfrak{g}$ and $Der_{(\alpha,\beta,\gamma,\tau)}\bar{\mathfrak{g}}, \forall(\alpha, \beta, \gamma, \tau) \in \mathbb{C}^4.$,*

Proof. Let $\mathfrak{g} = (V, \cdot)$ and $\bar{\mathfrak{g}} = (\bar{V}, *)$ be two Malcev algebras of the type Lie and let us consider $D \in Der_{(\alpha,\beta,\gamma,\tau)}\mathfrak{g}$, for any $(\alpha, \beta, \gamma, \tau) \in \mathbb{C}^4$ and for all $x, y, z \in \bar{\mathfrak{g}}$. Then,

$$\alpha D\left(f^{-1}(x) \cdot f^{-1}(y)\right) \cdot \left(f^{-1}(x) \cdot f^{-1}(z)\right) + \beta\left(f^{-1}(x) \cdot f^{-1}(y)\right) \cdot D\left(f^{-1}(x) \cdot f^{-1}(z)\right) =$$

$$\gamma D\left(\left(\left(f^{-1}(x) \cdot f^{-1}(z)\right) \cdot f^{-1}(y)\right) \cdot f^{-1}(x)\right) + \tau D\left(\left(\left(f^{-1}(z) \cdot f^{-1}(x)\right) \cdot f^{-1}(x)\right) \cdot f^{-1}(y)\right).$$

It is deduced that

$$\gamma D\left(\left(\left(f^{-1}(x)\cdot f^{-1}(z)\right)\cdot f^{-1}(y)\right)\cdot f^{-1}(x)\right)=\gamma D\left(\left(f^{-1}(x*z)\cdot f^{-1}(y)\right)\cdot f^{-1}(x)\right)=$$
$$\gamma Df^{-1}((x*z)*y)\cdot f^{-1}(x))=\gamma Df^{-1}(((x*z)*y)*x),$$

and, similarly,

$$\tau D\left(\left(\left(f^{-1}(z)\cdot f^{-1}(x)\right)\cdot f^{-1}(x)\right)\cdot f^{-1}(y)\right)=\tau Df^{-1}(((z*x)*x)*y)$$

$$\alpha D\left(f^{-1}(x)\cdot f^{-1}(y)\right)\cdot\left(f^{-1}(x)\cdot f^{-1}(z)\right)=\alpha Df^{-1}(x*y)\cdot f^{-1}(x*z)$$

$$\beta\left(f^{-1}(x)\cdot f^{-1}(y)\right)\cdot D\left(f^{-1}(x)\cdot f^{-1}(z)\right)=\beta f^{-1}(x*y)\cdot Df^{-1}(x*z).$$

Thus,

$$\alpha Df^{-1}(x*y)\cdot f^{-1}(x*z)+\beta f^{-1}(x*y)\cdot Df^{-1}(x*z)=\gamma Df^{-1}(((x*z)*y)*x)+\tau Df^{-1}(((z*x)*x)*y).$$

Now, the result of applying f to the previous expression is

$$\alpha\left(fDf^{-1}\right)(x*y)*(x*z)+\beta(x*y)*\left(fDf^{-1}\right)(x*z)=\gamma\left(fDf^{-1}\right)(((x*z)*y)*x)+\tau\left(fDf^{-1}\right)(((z*x)*x)*y).$$

Thus, $fDf^{-1}\in Der_{(\alpha,\beta,\gamma,\tau)}\bar{\mathfrak{g}}$, which concludes the proof. □

An immediate consequence of this result is the following.

Corollary 1. *Let $\mathfrak{g}$ be a Lie algebra. The dimension of the vector space $Der_{(\alpha,\beta,\gamma,\tau)}\mathfrak{g}$ is an invariant of the algebra, for all $(\alpha,\beta,\gamma,\tau)\in\mathbb{C}^4$.*

Lemma 1. *(Technical Lemma) Let d be a derivation of a Lie algebra $\mathfrak{g}$. The following expressions are verified*

1. $d[[[z,x],x],y]=d[[x,y],[x,z]]-d[[[y,z],x],x]$
2. $d[[[y,x],x],z]=d[[x,z],[x,y]]-d[[[z,y],x],x]$
3. $d[[[x,z],y],x]=d[[x,y],[x,z]]-d[[[z,x],x],y]$
4. $d[[[x,y],z],x]=d[[x,z],[x,y]]-d[[[y,x],x],z].$

Proof. All expressions are immediate consequences of the properties of the derivations (see Section 2). □

Lemma 2. *Let $\mathfrak{g}=(V,[,])$ be a Lie algebra. Then,*

$$Der_{(\alpha,\beta,\gamma,\tau)}\mathfrak{g}=Der_{(\alpha+\beta,\alpha+\beta,2\gamma,2\tau)}\mathfrak{g}\cap Der_{(\alpha-\beta,\beta-\alpha,0,0)}\mathfrak{g}$$

Proof. Suppose $D\in Der_{(\alpha,\beta,\gamma,\tau)}\mathfrak{g}$. Then, for all $(x,y,z)\in\mathfrak{g}$, we have

$$\alpha[d[x,y],[x,z]] + \beta[[x,y],d[x,z]] = \gamma d[[[x,z],y],x] + \tau d[[[z,x],x],y].$$

Charging now y and z between themselves, we have

$$\alpha[d[x,z],[x,y]] + \beta[[x,z],d[x,y]] = \gamma d[[[x,y],z],x] + \tau d[[[y,x],x],z]$$

and by adding the two first expressions of Lemma 1 and taking the anti-skew property of the Lie bracket into consideration, we have

$$(\alpha - \beta)[d[x,y],[x,z]] + (\beta - \alpha)[[x,y],d[x,z]] =$$
$$\gamma(d[[[x,z],y]x] + d[[[x,y],z]x]) + \tau(d[[[z,x],x],y] + d[[[y,x],x],z]).$$

Similarly, starting from the two last expressions of Lemma 1, we obtain

$$d[[[z,x],x],y] + d[[[y,x],x],z] = 0$$

and by repeating the same procedure we obtain

$$d[[[x,z],y],x] + d[[[x,y],z],x] = 0.$$

Now, starting from both expressions, we have

$$(\alpha - \beta)[d[x,y],[x,z]] + (\beta - \alpha)[[x,y],d[x,z]] = 0.$$

Therefore, $D \in Der_{(\alpha-\beta,\beta-\alpha,0,0)}\mathfrak{g}$.

Now, by subtracting the two first expressions of the proof and taking into account the anti-skew property, we have $(\alpha + \beta)[d[x,y],[x,z]] + (\beta + \alpha)[[x,y],d[x,z]] = \gamma(d[[[x,z],y],x] - d[[[x,y],z],x]) + \tau(d[[[z,x],x],y] - d[[[y,x],x],z])$.

We use now in the previous equality the two expressions $d[[[y,x],x],z] = -d[[[z,x],x],y]$ and $d[[[x,y],z],x] = -d[[[x,z],y],x]$, respectively, obtained from previous expressions.

We have that $(\alpha + \beta)[d[x,y],[x,z]] + (\alpha + \beta)[[x,y],d[x,z]] = 2\gamma d[[[x,z],y],x] + 2\tau d[[[z,x],x],y]$. It involves that $D \in Der_{(\alpha+\beta,\alpha+\beta,2\gamma,2\tau)}\mathfrak{g}$. Therefore, it is verified that $Der_{(\alpha,\beta,\gamma,\tau)}\mathfrak{g} \subset Der_{(\alpha+\beta,\alpha+\beta,2\gamma,2\tau)}\mathfrak{g} \cap Der_{(\alpha-\beta,\beta-\alpha,0,0)}\mathfrak{g}$.

If $D \in Der_{(\alpha+\beta,\alpha+\beta,2\gamma,2\tau)}\mathfrak{g} \cap Der_{(\alpha-\beta,\beta-\alpha,0,0)}\mathfrak{g}$, then D verifies both equations $(\alpha + \beta)[d[x,y],[x,z]] + (\alpha + \beta)[[x,y],d[x,z]] = 2\gamma d[[[x,z],y],x] + 2\tau d[[[z,x],x],y]$ and $(\alpha - \beta)[d[x,y],[x,z]] + (\beta - \alpha)[[x,y],d[x,z]] = 0$.

Then, by adding these last equations and simplifying, we observe that D verifies

$$\alpha[d[x,y],[x,z]] + \beta[[x,y],d[x,z]] = \gamma d[[[x,z],y],x] + \tau d[[[z,x],x],y].$$

Thus, $D \in Der_{(\alpha,\beta,\gamma,\tau)}\mathfrak{g} = Der_{(\alpha+\beta,\alpha+\beta,2\gamma,2\tau)}\mathfrak{g} \cap Der_{(\alpha-\beta,\beta-\alpha,0,0)}\mathfrak{g}$.

Therefore, $Der_{(\alpha,\beta,\gamma,\tau)}\mathfrak{g} = Der_{(\alpha+\beta,\alpha+\beta,2\gamma,2\tau)}\mathfrak{g} \cap Der_{(\alpha-\beta,\beta-\alpha,0,0)}\mathfrak{g}$, which completes the proof. $\square$

Theorem 2. *Let $\mathfrak{g}$ be a Lie algebra. Then, for all $(\alpha, \beta, \gamma, \tau) \in \mathbb{C}^4$, it exists $(\lambda_1, \lambda_2) \in \mathbb{C}^2$ such that $Der_{(\alpha,\beta,\gamma,\tau)}\mathfrak{g} \subset \mathbb{C}^2$ is one of the following four sets: $Der_{(0,0,\lambda_1,\lambda_2)}\mathfrak{g}$; $Der_{(1,-1,\lambda_1,\lambda_2)}\mathfrak{g}$; $Der_{(1,0,\lambda_1,\lambda_2)}\mathfrak{g}$; or $Der_{(1,1,\lambda_1,\lambda_2)}\mathfrak{g}$.*

Proof. Consider $(\alpha, \beta, \gamma, \tau) \in \mathbb{C}^4$. We distinguish the following cases

Case 1: $\alpha + \beta = 0$. We distinguish now the following two subcases:

1.1 $\alpha = \beta = 0$. Then, $Der_{(\alpha,\beta,\gamma,\tau)}\mathfrak{g} = Der_{(0,0,\gamma,\tau)}\mathfrak{g}$. Therefore, $\gamma = \lambda_1$ and $\lambda_2 = \tau$.

1.2 $\alpha = -\beta$. In this subcase, by Lemma 2, we have that

$$Der_{(\alpha,\beta,\gamma,\tau)}\mathfrak{g} = Der_{(0,0,2\gamma,2\tau)}\mathfrak{g} \cap Der_{(-2\beta,2\beta,0,0)}\mathfrak{g} = Der_{(0,0,\gamma,\tau)}\mathfrak{g} \cap Der_{(-1,1,0,0)}\mathfrak{g}.$$

Apart from that, it is also verified that

$$Der_{(-1,1,\gamma,\tau)}\mathfrak{g} = Der_{(0,0,2\gamma,2\tau)}\mathfrak{g} \cap Der_{(-2,2,0,0)}\mathfrak{g} = Der_{(0,0,\gamma,\tau)}\mathfrak{g} \cap Der_{(-1,1,0,0)}\mathfrak{g}.$$

Therefore, $Der_{(\alpha,\beta,\gamma,\tau)}\mathfrak{g} = Der_{(-1,1,\gamma,\tau)}\mathfrak{g}$. It involves that $\lambda_1 = \gamma$ and $\lambda_2 = \tau$.

Case 2: $\alpha + \beta \neq 0$. Two subcases are also considered:

2.1 $\alpha \neq \beta$.

By Lemma 2, we have $Der_{(\alpha,\beta,\gamma,\tau)}\mathfrak{g} = Der_{(1,1,\frac{2\gamma}{\alpha+\beta},\frac{2\tau}{\alpha+\beta})}\mathfrak{g} \cap Der_{(1,-1,0,0)}\mathfrak{g}$.

Since $Der_{(1,0,\frac{\gamma}{\alpha+\beta},\frac{\tau}{\alpha+\beta})}\mathfrak{g} = Der_{(1,1,\frac{2\gamma}{\alpha+\beta},\frac{2\tau}{\alpha+\beta})}\mathfrak{g} \cap Der_{(1,-1,0,0)}\mathfrak{g}$, it is deduced that $Der_{(\alpha,\beta,\gamma,\tau)}\mathfrak{g} = Der_{(1,0,\frac{\gamma}{\alpha+\beta},\frac{\tau}{\alpha+\beta})}\mathfrak{g}$. It involves that $\lambda_1 = \frac{\gamma}{\alpha+\beta}$ and $\lambda_2 = \frac{\tau}{\alpha+\beta}$.

2.2 $\alpha = \beta$.

In this subcase, $Der_{(\alpha,\beta,\gamma,\tau)}\mathfrak{g} = Der_{(1,1,\frac{\gamma}{\alpha},\frac{\tau}{\alpha})}\mathfrak{g}$. Therefore, $\lambda_1 = \frac{\gamma}{\alpha}$ and $\lambda_2 = \frac{\tau}{\alpha}$.

□

These two two-parameter sets $Der_{(1,0,\lambda_1,\lambda_2)}\mathfrak{g}$ and $Der_{(1,1,\lambda_1,\lambda_2)}\mathfrak{g}$ previously defined allow us to define the following invariant two-parameter functions of Lie algebras.

Definition 2. *The functions $\bar{\psi}_{\mathfrak{g}}, \bar{\psi}_{\mathfrak{g}}^0 : \mathbb{C}^2 \mapsto \mathbb{N}$ defined, respectively, as $(\bar{\psi}_{\mathfrak{g}})(\alpha,\beta) = \dim Der_{(1,1,\alpha,\beta)}\mathfrak{g}$ and $(\bar{\psi}_{\mathfrak{g}}^0)(\alpha,\beta) = \dim Der_{(1,0,\alpha,\beta)}\mathfrak{g}$ are called $\bar{\psi}_{\mathfrak{g}}$ and $\bar{\psi}_{\mathfrak{g}}^0$ invariant functions corresponding to the $(\alpha,\beta,\gamma,\tau)$-derivations of $\mathfrak{g}$.*

Corollary 2. *If two Malcev algebras of the type Lie $\mathfrak{g}$ and $\mathfrak{f}$ are isormorphic, then $\bar{\psi}_{\mathfrak{g}} = \bar{\psi}_{\mathfrak{f}}$ and $\bar{\psi}_{\mathfrak{g}}^0 = \bar{\psi}_{\mathfrak{f}}^0$.*

Note that the function $\bar{\psi}$ is a two-parameter function, whereas the function ψ by Novotný and Hrivnák [7] is one-parameter. It implies that both functions are structurally different. However, it can be thought that ψ could be obtained as a particular case of $\bar{\psi}$ by simply taking one of the parameters as a constant. The following counter-example shows that it is not possible.

Indeed, we now compare the function $\bar{\psi}$ with the invariant function ψ and prove that both functions are totally different. To do this, we compute both functions for a same Lie algebra, in the particular case of being $\alpha = 1$. Concretely, we use the Lie algebra induced by the Lorentz group $SO(3,1)$, which we denote by $\mathfrak{g}_6$.

Computing $\psi_{\mathfrak{g}_6}$, for $\alpha = 1$

Let us recall that Minkowski defined the spacetime as a four-dimensional manifold with the metric $ds^2 = -c^2dt^2 + dx^2 + dy^2 + dz^2$. We introduce the metric tensor

$$\eta = \begin{bmatrix} -1 & 0 & 0 & 0 \\ 0 & 1 & 0 & 0 \\ 0 & 0 & 1 & 0 \\ 0 & 0 & 0 & 1 \end{bmatrix}.$$

If we rename $(ct, x, y, z) \rightarrow (x^0, x^1, x^2, x^3)$, then the expression ds^2 can be written as $ds^2 = \eta_{\mu\gamma} dx^\mu dx^\gamma$ (summed over μ and γ). Recall that this distance is invariant under the following type of transformations $x^\mu \rightarrow \lambda^\mu_\gamma x^\gamma$ such that the coefficients λ^μ_γ are the elements of a matrix Λ (which is called *Lorentz transformations*) that satisfies $\Lambda^t \eta \Lambda = \eta$. Since the metric in the three-dimensional Euclidean space corresponds to the identity matrix, if R is the matrix of a rotation, then $R^t 1 R = 1$ and comparing this expression with $\Lambda^t \eta \Lambda = \eta$ it is possible to say that the Lorentz transformations are rotations in the Minkowski space. These transformations form a group called the *Lorentz group* $SO(3,1)$.

Now, we focus our study on the *infinitesimal* Lorentz transformations. A Lorentz transformation matrix can be written as $\Lambda^\mu_\gamma = \delta^\mu_\gamma + \lambda^\mu_\gamma$, where the parameters λ^μ_γ are infinitesimal and verify that $\lambda^\mu_\gamma = -\lambda^\gamma_\mu$ so that the Lorentz transformation is valid. The action of this transformation on the coordinates x^μ in the Minkowski space can be written as $\delta x^\mu = \Lambda^\mu_\gamma x^\gamma$.

If we define $A_{\rho\sigma}$ such that $\Lambda^\mu_\gamma = \frac{1}{2}\lambda^{\rho\sigma}(A_{\rho\sigma})^\mu_\gamma$, we can write the above action as $\delta x^\mu = \frac{1}{2}\lambda^{\rho\sigma}(A_{\rho\sigma})^\mu_\gamma x^\gamma$. Then, it is easily proved that $(A_{\rho\sigma})^\mu_\gamma = \delta^\mu_\rho \eta_{\sigma\gamma} - \delta^\mu_\sigma \eta_{\rho\gamma}$.

Explicitly,

$$A_{10} = \begin{pmatrix} 0 & -1 & 0 & 0 \\ -1 & 0 & 0 & 0 \\ 0 & 0 & 0 & 0 \\ 0 & 0 & 0 & 0 \end{pmatrix} \quad A_{20} = \begin{pmatrix} 0 & 0 & -1 & 0 \\ 0 & 0 & 0 & 0 \\ -1 & 0 & 0 & 0 \\ 0 & 0 & 0 & 0 \end{pmatrix} \quad A_{30} = \begin{pmatrix} 0 & 0 & 0 & -1 \\ 0 & 0 & 0 & 0 \\ 0 & 0 & 0 & 0 \\ -1 & 0 & 0 & 0 \end{pmatrix}$$

$$A_{12} = \begin{pmatrix} 0 & 0 & 0 & 0 \\ 0 & 0 & 1 & 0 \\ 0 & -1 & 0 & 0 \\ 0 & 0 & 0 & 0 \end{pmatrix} \quad A_{23} = \begin{pmatrix} 0 & 0 & 0 & 0 \\ 0 & 0 & 0 & 0 \\ 0 & 0 & 0 & 1 \\ 0 & 0 & -1 & 0 \end{pmatrix} \quad A_{31} = \begin{pmatrix} 0 & 0 & 0 & 0 \\ 0 & 0 & 0 & -1 \\ 0 & 0 & 0 & 0 \\ 0 & 1 & 0 & 0 \end{pmatrix}$$

Now, by defining the Lie product as the usual commutator $[A_{ij}, A_{hk}] = A_{ij} \cdot A_{hk} - A_{hk} \cdot A_{ij}$, $A_{10}, A_{20}, A_{30}, A_{12}, A_{23}$ and A_{31} generate a Lie algebra, which we denote by $\mathfrak{g}_6$.

Let us consider $d \in Der_{(1,1,1,1)}\mathfrak{g}_6$ and let $A = (a_{ij})$, $1 \le i, j \le 6$ be the 6×6 square matrix associated with the endomorphism d.

To obtain the elements of this matrix, for the pair of generators (e_i, e_j), with $i < j$, the derivation d satisfies $d([e_i, e_j]) = [d(e_i), e_j] + [e_i, d(e_j)]$ and $d(e_i) = \sum_{h=1}^6 a_{ih} e_h$. In this way, the following conditions are obtained. This can be seen in the following Table 1.

Table 1. Condition obtained.

From Pair (e_i, e_j)	Conditions
(e_1, e_2)	$a_{41} = a_{14}$, $a_{42} = a_{24}$, $a_{43} = -a_{15} - a_{26}$, $a_{44} = a_{11} + a_{22}$, $a_{45} = -a_{13}$, $a_{46} = -a_{23}$.
(e_1, e_3)	$a_{61} = a_{16}$, $a_{62} = -a_{15} - a_{34}$, $a_{63} = a_{36}$, $a_{64} = -a_{32}$, $a_{65} = -a_{12}$, $a_{66} = a_{33} + a_{11}$.
(e_1, e_4)	$a_{21} = -a_{12}$, $a_{22} = a_{11} + a_{44}$, $a_{23} = -a_{46}$, $a_{24} = a_{42}$, $a_{25} = -a_{16}$, $a_{26} = a_{15} - a_{43}$.
(e_1, e_5)	$a_{13} = 0$, $a_{54} = 0$, $a_{12} - a_{56} = 0$, $a_{16} + a_{52} = 0$, $a_{14} + a_{53} = 0$.
(e_1, e_6)	$a_{31} = -a_{13}$, $a_{32} = -a_{64}$, $a_{33} = a_{11} + a_{66}$, $a_{34} = a_{15} - a_{62}$, $a_{35} = -a_{14}$, $a_{36} = a_{63}$.
(e_2, e_3)	$a_{51} = -a_{26} - a_{34}$, $a_{52} = a_{25}$, $a_{53} = a_{35}$, $a_{54} = -a_{31}$, $a_{55} = a_{22} + a_{33}$, $a_{56} = -a_{21}$.
(e_2, e_4)	$a_{11} = a_{22} + a_{44}$, $a_{12} = -a_{21}$, $a_{13} = -a_{45}$, $a_{14} = a_{41}$, $a_{15} = a_{26} - a_{43}$, $a_{16} = -a_{25}$.
(e_2, e_5)	$a_{31} = -a_{54}$, $a_{32} = -a_{23}$, $a_{33} = a_{22} + a_{55}$, $a_{34} = a_{26} - a_{51}$, $a_{35} = a_{53}$, $a_{36} = -a_{24}$.
(e_2, e_6)	$a_{23} - a_{64} = 0$, $-a_{21} + a_{65} = 0$, $a_{25} + a_{61} = 0$, $a_{24} + a_{63} = 0$.
(e_3, e_5)	$a_{21} = -a_{56}$, $a_{22} = a_{33} + a_{55}$, $a_{23} = -a_{32}$, $a_{24} = -a_{36}$, $a_{25} = a_{52}$, $a_{26} = a_{34} - a_{51}$.
(e_3, e_6)	$a_{11} = a_{33} + a_{66}$, $a_{12} = -a_{65}$, $a_{13} = -a_{31}$, $a_{14} = -a_{35} + a_{34}$, $a_{15} = -a_{62}$, $a_{16} = a_{61}$.
(e_4, e_5)	$a_{61} = -a_{52}$, $a_{62} = a_{43} + a_{51}$, $a_{63} = -a_{42}$, $a_{64} = -a_{46}$, $a_{65} = -a_{56}$, $a_{66} = a_{44} + a_{55}$.
(e_4, e_6)	$a_{51} = a_{43} + a_{62}$, $a_{52} = -a_{61}$, $a_{53} = -a_{41}$, $a_{54} = -a_{45}$, $a_{55} = a_{44} + a_{66}$, $a_{56} = -a_{65}$.
(e_5, e_6)	$a_{41} = -a_{53}$, $a_{42} = -a_{63}$, $a_{43} = a_{51} + a_{62}$, $a_{44} = a_{55} + a_{66}$, $a_{45} = -a_{54}$, $a_{46} = -a_{64}$.
(e_3, e_4)	$-a_{32} + a_{46} = 0$, $a_{31} - a_{45} = 0$, $a_{36} + a_{42} = 0$, $a_{35} + a_{41} = 0$.

From these conditions on a_{ij} and $\forall\, a_{41}, a_{42}, a_{44}, a_{46}, a_{55}, a_{61}, a_{65}, a_{66} \in \mathbb{C}$, we have the following conditions shown in Table 2.

Table 2. Conditions obtained.

$a_{11} = a_{55}$,	$a_{12} = -a_{65}$,	$a_{13} = 0$,	$a_{14} = a_{41}$,	$a_{15} = 0$,	$a_{16} = a_{61}$.
$a_{21} = a_{65}$,	$a_{22} = a_{66}$,	$a_{23} = -a_{46}$,	$a_{24} = a_{42}$,	$a_{25} = -a_{61}$,	$a_{26} = 0$.
$a_{31} = 0$,	$a_{32} = a_{46}$,	$a_{33} = a_{44}$,	$a_{34} = 0$,	$a_{35} = -a_{41}$,	$a_{36} = -a_{42}$,
		$a_{43} = 0$,		$a_{45} = 0$.	
$a_{51} = 0$,	$a_{52} = -a_{61}$,	$a_{53} = -a_{41}$,	$a_{54} = 0$,		$a_{56} = -a_{65}$.
	$a_{62} = 0$,	$a_{63} = -a_{42}$,	$a_{64} = -a_{46}$.		

This implies that $\psi_{\mathfrak{g}_6}(1) = dim\left(Der_{(1,1,1,1)}\mathfrak{g}_6\right) = 8$.

Computing $\bar{\psi}_{\mathfrak{g}_6}$, for $\alpha = 1$

Let us consider $d \in Der_{(1,1,1,1)}\mathfrak{g}_6$. Then, $[d[u,v],[u,w]] + [[u,v],d[u,w]] = d[[u,w],v],u] + d[[[w,u],u],v]$, $\forall u, v, w \in \mathfrak{g}_6$.

To obtain the elements a_{ij} of the corresponding 6×6 square matrix associated with d, we see that for each triplets of generators (e_i, e_j, e_k) of the algebra, the previous expression is written as

$$[d[e_i, e_j], [e_i, e_k]] + [[e_i, e_j], d[e_i, e_k]] = d[[[e_i, e_k], e_j], e_i] + d[[[e_k, e_i], e_i], e_j].$$

Starting from it, we obtain the following conditions shown in Table 3.

Table 3. Conditions obtained.

From Triplet (e_i, e_j, e_k)	Conditions
(e_1, e_2, e_3)	$a_{51} = a_{43} + a_{62},\ \ a_{52} = -a_{61},\ \ a_{53} = -a_{41},$ $a_{54} = -a_{45},\ \ a_{55} = a_{66} + a_{44},\ \ a_{56} = -a_{65}.$
(e_1, e_2, e_4)	$a_{11} = a_{22} + a_{44},\ \ a_{12} = -a_{21},\ \ a_{13} = -a_{45},$ $a_{14} = -a_{41},\ \ a_{15} = a_{26} - a_{43},\ \ a_{16} = -a_{25}.$
(e_1, e_2, e_5)	$0 = 0$
(e_1, e_2, e_6)	$a_{32} + a_{46} = 0,\ \ e_{31} + a_{45} = 0,\ \ a_{36} + a_{42} = 0,$ $a_{35} + a_{41} = 0.$
(e_1, e_3, e_4)	$-a_{23} + a_{64} = 0,\ \ a_{21} - a_{65} = 0,\ \ a_{25} + a_{61} = 0$ $a_{24} + a_{63} = 0.$
(e_1, e_3, e_5)	$0 = 0$
(e_1, e_3, e_6)	$a_{33} + a_{66} = a_{11},\ \ a_{65} = -a_{12},\ \ a_{31} = -a_{13},$ $a_{35} = -a_{14},\ \ a_{34} - a_{62} = a_{15},\ \ a_{61} = a_{16}.$
(e_1, e_4, e_5)	$0 = 0$
(e_1, e_4, e_6)	$a_{51} = -a_{26} - a_{34},\ \ a_{52} = a_{25},\ \ a_{53} = a_{35},$ $a_{54} = -a_{31},\ \ a_{55} = a_{22} + a_{33},\ \ a_{56} = -a_{21}.$
(e_1, e_5, e_6)	$0 = 0$
(e_2, e_3, e_4)	$-a_{13} + a_{54} = 0,\ \ a_{12} - a_{56} = 0,\ \ a_{16} + a_{52} = 0,$ $a_{14} + a_{53} = 0.$
(e_2, e_3, e_5)	$a_{21} = -a_{56},\ \ a_{22} = a_{33} + a_{55},\ \ a_{23} = -a_{32},$ $a_{24} = -a_{36},\ \ a_{25} = a_{52},\ \ a_{26} = a_{34} - a_{51}.$
(e_2, e_3, e_6)	$0 = 0$
(e_2, e_4, e_5)	$a_{61} = -a_{16},\ \ a_{62} = a_{15} + a_{34},\ \ a_{63} = -a_{36},$ $a_{64} = a_{32},\ \ a_{65} = a_{12},\ \ a_{66} = -a_{11} - a_{33}.$
(e_2, e_4, e_6)	$0 = 0$
(e_2, e_5, e_6)	$0 = 0$
(e_3, e_4, e_5)	$0 = 0$
(e_3, e_4, e_6)	$0 = 0$
(e_4, e_5, e_6)	$a_{41} = -a_{53},\ \ a_{42} = -a_{63},\ \ a_{43} = a_{51} + a_{62},$ $a_{44} = a_{55} + a_{66},\ \ a_{45} = -a_{54},\ \ a_{46} = -a_{64}.$
(e_3, e_5, e_6)	$a_{41} = a_{14},\ \ a_{42} = a_{24},\ \ a_{43} = -a_{15} - a_{26},$ $a_{44} = a_{11} + a_{22},\ \ a_{45} = -a_{13},\ \ a_{46} = -a_{23}.$

It follows from these conditions for a_{ij} that $a_{ij} = 0,\ \ \forall i, j \in \{1, 2, 3, 4, 5, 6\}$. This implies that $\tilde{\psi}_{\mathfrak{g}_6}(1, 1) = dim\left(Der_{(1,1,1,1)}\mathfrak{g}_6\right) = 0$, which proves that $\psi \neq \tilde{\psi}$ in general.

3.2. The Quantum-Mechanical Model Based on a 5th Heisenberg Algebra

In this section, and by using the invariant function previously introduced $\bar{\psi}$, we prove the following result.

Theorem 3. Main Theorem

The five-dimensional classical-mechanical model built upon certain types of five-dimensional Lie algebras cannot be obtained as a limit process of a quantum-mechanical model based on a fifth Heisenberg algebra.

Proof. Let $\mathbb{H}_5$ be the fifth Heisenberg algebra generated by $\{e_1, \ldots, e_5\}$ and defined by the brackets $[e_1, e_3] = e_5$ and $[e_2, e_4] = e_5$.

Let us consider $d \in Der_{(1,1,1,1)}\mathbb{H}_5$. Then, $[d[u,v],[u,w]] + [[u,v],d[u,w]] = d[[[u,w],v],u] + d[[[w,u],u],v]$, $\forall u, v, w \in \mathbb{H}_5$.

To obtain the elements a_{ij} of the corresponding 5×5 square matrix associated with d, we see that for each triplet of generators (e_i, e_j, e_k) of the algebra, the previous expression is written as

$$[d[e_i, e_j], [e_i, e_k]] + [[e_i, e_j], d[e_i, e_k]] = d[[[e_i, e_k], e_j], e_i] + d[[[e_k, e_i], e_i], e_j].$$

Note that, in this case, there is no restriction on the elements of the matrix associated with d and, thus, $\bar{\psi}_{\mathbb{H}_5}(1,1) = dim\left(Der_{(1,1,1,1)}\mathbb{H}_5\right) = 25$.

For another part, let $\mathfrak{f}_5$ be the five-dimensional filiform Lie algebra, defined by $[e_1, e_3] = e_2$, $[e_1, e_4] = e_3$ and $[e_1, e_5] = e_4$.

Let us consider $d \in Der_{(1,1,1,1)}\mathfrak{f}_5$. Then, it is verified that $[d[u,v],[u,w]] + [[u,v],d[u,w]] = d[[[u,w],v],u] + d[[[w,u],u],v]$, $\forall u, v, w \in \mathfrak{f}_5$.

Similar to the previous case, to obtain the elements a_{ij} of the corresponding 5×5 square matrix associated with d, we see that, for each triplet of generators (e_i, e_j, e_k) of the algebra, the previous expression is written as

$$[d[e_i, e_j], [e_i, e_k]] + [[e_i, e_j], d[e_i, e_k]] = d[[[e_i, e_k], e_j], e_i] + d[[[e_k, e_i], e_i], e_j].$$

In this case, the restrictions of the matrix associated with d are $a_{21} = 0$, obtained from the bracket (e_1, e_3, e_5) and $a_{31} = a_{41}$ from (e_1, e_4, e_5), therefore $\bar{\psi}(1,1) = 23$.

Next, we use the highly non-trivial result, which was originally proved by Borel [17]: *If $\mathfrak{g}_0$ is a proper contraction of a complex Lie algebra $\mathfrak{g}$, then it holds:* $dim\left(Der\mathfrak{g}\right) < dim\left(Der\mathfrak{g}_0\right)$.

Indeed, according to Proposition 1 we obtain that

$$\bar{\psi}_{\mathbb{H}_5}(1,1) = dim\left(Der_{(1,1,1,1)}\mathbb{H}_5\right) = dim\left(Der\left(\mathbb{H}_5\right)\right) = 25$$

and

$$\bar{\psi}_{\mathfrak{f}_5}(1,1) = dim\left(Der_{(1,1,1,1)}\mathfrak{f}_5\right) = dim\left(Der\left(\mathfrak{f}_5\right)\right) = 23.$$

It implies that no proper contraction transforming the Heisemberg algebra $\mathbb{H}_5$ into the filiform Lie algebra $\mathfrak{f}_5$ exists. Thus, since both algebras are not isomorphic, the five-dimensional classical-mechanical model built upon a five-dimensional filiform Lie algebra cannot be obtained as a limit process of a quantum-mechanical model based on a fifth Heisenberg algebra. $\square$

4. Discussion and Conclusions

In this paper, we introduce an invariant two-parameter function of algebras, $\bar{\psi}$, and we have used it as a tool to study contractions of certain particular types of algebras.

Indeed, by means of this function, we have proved that there is no proper contraction between a fifth Heisenberg algebra and a filiform Lie algebra of dimension 5. It implies, as a main result, that the five-dimensional classical-mechanical model built upon a five-dimensional filiform Lie algebra cannot be obtained as a limit process of a quantum-mechanical model based on a fifth Heisenberg algebra.

We have also computed this function in the case of other types of algebras, for instance, Malcev algebras of the type Lie and the Lie algebra induced by the Lorentz group $SO(3,1)$.

Apart from continuing this study with with higher-dimensional algebras, we indicate next some open problems to be dealt with in future work, most of them with the objective of trying to find some possible interesting physical applications for the filiform Lie algebras. They are the following

1. As mentioned above, in 2007, Hrivnák and Novotný introduced the invariant functions ψ and φ as a tool to study contractions of Lie algebras [7]. Those are one-parameter functions. We have now defined the two-parameter invariant function $\bar{\psi}$. It would be good to search new invariant functions to continue with this research, for instance, some related with twisted cocycles of Lie algebras.

2. It would also be good to find necessary and sufficient conditions which characterize contractions of Lie algebras.

3. One of the possible physical applications of the present topic is given by the possibility of describing a many-body system based on interacting spinless boson particles located in a lattice of n sites by means of a filiform Lie algebra. This system could be a kind of Bose–Hubbard model, which is well known in the condensed matter community and widely studied. The Hamiltonian corresponding to that system can be described in terms of semi-simple Lie algebras and is a quadratic model since it contains up to two-body operators. Therefore, we wonder if we could describe the same system employing filiform Lie algebras and if we could obtain new information using the tools developed in this manuscript.

    To perform this task, it is necessary to write the boson operators involved in the Hamiltonian in term of new ones that fulfill the commutation relations for a given filiform Lie algebra. However, at that point, we find the difficulty that we should employ a tensorial product of two filiform Lie algebras in order to describe the system properly. That means that an isomorphism between the semi-simple Lie algebra of the original hamiltonian and the filiform Lie algebra proposed to describe the physical system should exist. Fortunately, it seems that we have obtained a theorem that can confirm that kind of isomorphism.

    Now, the advantage that we gain employing a filiform Lie algebra instead of a semi-simple Lie algebra is that we could map a non-linear problem such as the problem described by a system with up to two-body interactions onto a linear problem with just one-body interactions. On the other hand, once we have described the system in terms of the filiform Lie algebra, it is necessary to define the branching rules, that is to find the irreducible representations of an algebra $\mathfrak{g}'$ contained in a given representation of $\mathfrak{g}$. Since the representations are interpreted as quantum mechanical states, it is necessary to provide a complete set of quantum numbers (labels) to characterize uniquely the basis of the system. This is a non-trivial task that it may even lead to a further research.

4. Another possible physical applications of the present topic is to study phase spaces by using filiform Lie algebras as a tool.

    In this respect, Arzano and Nettel [18] in 2016 introduced a general framework for describing deformed phase spaces with group valued momenta. Using techniques from the theory of Poisson–Lie groups and Lie bialgebras, they developed tools for constructing Poisson structures on the deformed phase space starting from the minimal input of the algebraic structure of the generators of the momentum Lie group. These tools developed are used to derive Poisson structures on examples of group momentum space much studied in the literature such as the n-dimensional generalization of the κ-deformed momentum space and the $SL(2, R)$ momentum space in three space-time dimensions. They also discussed classical momentum observables associated to multiparticle systems and argued that these combined according the usual four-vector addition despite the non-Abelian group structure of momentum space (see [18] for further information).

In that paper, the authors work with a phase space $\Gamma = T \times G$, given by the Cartesian product of a n-dimensional Lie group configuration space T and a n-dimensional Lie group momentum space G. Since T and G are Lie groups, we can consider their associated Lie algebras $\mathfrak{t}$ and $\mathfrak{g}$ so that we can define a Lie–Poisson algebra, which can endow a mathematical structure to the phase space Γ. Indeed, Arzano and Nettel considered a phase space Γ in which the component related to momentum is an n-dimensional Lie sub-group of the $(n + 2)$-dimensional Lorentz group $SO(n + 1, 1)$, denoted as $AN(n)$.

Taking into consideration this paper, we have tried to construct a phase space similar to the one by those authors, although we have taken the $(n + 2)$-dimensional Lorentz group $SO(n + 1, 2)$ as the Lie group related to momentum.

We began our research on this subject considering the Lie group $SO(2, 2)$ and using the same procedure as Arzano and Nettel did. However, we realized that that attempt was going to be very complicated because of the great dimensions of the matrices involved (in the computations, a 49×49 $r-$matrix appeared).

Therefore, the fact of finding a Poisson structure that allows us to endow the phase space $\Gamma = T \times SO(n + 1, 2)$ with a mathematical structure is another problem, which we consider open.

5. Finally, semi-invariant functions of algebras could also be considered to study contractions of Lie Algebras (see [19], for instance).

We will dedicate our efforts to these objectives in future work.

5. Materials and Methods

Since this is a work on pure and applied mathematics, no type of materials different from the usual ones in a theoretical investigation was needed. Indeed, on the one hand, only the existing bibliography on the subject and, on the other hand, a suitable symbolic computation package were used. In the same way, with regard to the methodology used for the writing of the manuscript, it was also the usual one in research work of this nature, namely, based on already established hypotheses and known results.

We used the SAGE symbolic computation package for computations. SageMath, which is a free open-source mathematics software system licensed under the GPL, builds on top of many existing open-source packages, such as matplotlib, Sympy, Maxima, GAP, R and many more (see [20], for instance).

Author Contributions: All of the authors wrote the paper and J.N.-V. and P.P.-F. checked the proofs.

Funding: This research was funded by the Spanish Ministerio de Ciencia e Innovación and Junta de Andalucía via grants No. MTM2013-40455-P and No. FQM-326 (J.N.-V.) and No. FQM-160 (P.P.-F.).

Acknowledgments: The authors gratefully acknowledge the financial support above mentioned.

Conflicts of Interest: The authors declare no conflict of interest.

References

1. Inönü, E.; Wigner, E. On the contraction of groups and their representations. *Proc. Nat. Acad. Sci. USA* **1953**, *39*, 510–524. [CrossRef] [PubMed]

2. Inönü, E.; Wigner, E. On a particular type of convergence to a singular matrix. *Proc. Nat. Acad. Sci. USA* **1954**, *40*, 119–121. [CrossRef] [PubMed]

3. Doebner, H.D.; Melsheimer O. On a class of generalized group contractions. *Nuovo Cimento A* **1967**, *49*, 306–311. [CrossRef]

4. Burde, D. Degenerations of nilpotent Lie algebras. *J. Lie Theory* **1999**, *9*, 193–202.

5. Burde, D. Degenerations of 7-dimensional nilpotent Lie algebras. *Commun. Algebra* **2005**, *33*, 1259–1277. [CrossRef]

6. Steinhoff, C. Klassifikation und Degeneration von Lie Algebren. Ph.D. Thesis, Universität Düsseldorf, Düsseldorf, Germany, 1997.

7. Novotný, P.; Hrivnák, J. On (α, β, γ)-derivations of Lie algebras and corresponding invariant functions. *J. Geometry Phys.* **2008**, *58*, 208–217. [CrossRef]

8. Escobar, J.M.; Núñez, J.; Pérez-Fernández, P. On contractions of Lie algebras. *Math. Comput. Sci.* **2016**. [CrossRef]

9. Humphreys, J.E. *Introduction to Lie Algebras and Representation Theory*; Springer: New York, NY, USA, 1972.

10. Vergne, M. Cohomologie des algèbres de Lie nilpotentes. Application à l'étude de la variété des algebres de Lie nilpotentes. *Bull. Soc. Math. France* **1970**, *98*, 81–116. [CrossRef]

11. Sagle, A.A. Malcev Algebras. *Trans. Am. Math. Soc.* **1961**, *101*, 426–458. [CrossRef]

12. Dirac, P.A.M. *Lectures on Quantum Mechanics*; Yeshiva University: New York, NY, USA, 1964.

13. Lipkin, H.J.; Weisberger, W.I.; Peshkin, M. Magnetic Charge Quantization and Angular Momentum. *Ann. Phys.* **1969**, *53*, 203–214. [CrossRef]

14. Günaydin, M. Exceptionality, supersymmetry and non-associativity in Physics. In Proceedings of the Bruno Zumino Memorial Meeting, Geneva, Switzerland, 27–28 April 2015.

15. Günaydin, M.; Minic, D. Nonassociativity, Malcev algebras and string theory. *Fortschritte der Physik* **2013**, *61*, 873–892.

16. Falcón, O.J.; Falcón, R.M.; Núñez, J. A computational algebraic geometry approach to enumerate Malcev magma algebras over finite fields. *Math. Methods Appl. Sci.* **2016**, *39*, 4901–4913. [CrossRef]

17. Borel, A. *Linear Algebraic Groups*; Benjamin, Inc.: New York, NY, USA, 1969.

18. Arzano, M.; Nettel, F. Deformed phase spaces with group valued momenta. *Phys. Rev. D* **2016**, *94*, 085004. [CrossRef]

19. Nesterenko, M.; Popovych, R. Contractions of Low-Dimensional Lie Algebras. *J. Math. Phys.* **2006**, *47*, 123515. [CrossRef]

20. On SAGE symbolic computation package. Available online: http://www.sagemath.org/ (accessed on 11 October 2019).

Mathematical and Computational Applications

Article

Solving Nonholonomic Systems with the Tau Method

Alexandra Gavina [1,*], José M. A. Matos [1,2] and Paulo B. Vasconcelos [2,3]

1   Laboratório Engenharia Matemática, Instituto Superior Engenharia Porto, R. Dr. António Bernardino de Almeida 431, 4200-072 Porto, Portugal; jma@isep.ipp.pt
2   Centro de Matemática, Universidade do Porto, Rua do Campo Alegre 687, 4169-007 Porto, Portugal; pjv@fe.up.pt
3   Faculdade de Economia, Universidade do Porto, Rua Dr. Roberto Frias, s/n, 4200-464 Porto, Portugal
*   Correspondence: alg@isep.ipp.pt

Received: 27 September 2019; Accepted: 17 October 2019; Published: 19 October 2019

Abstract: A numerical procedure based on the spectral Tau method to solve nonholonomic systems is provided. Nonholonomic systems are characterized as systems with constraints imposed on the motion. The dynamics is described by a system of differential equations involving control functions and several problems that arise from nonholonomic systems can be formulated as optimal control problems. Applying the Pontryagins maximum principle, the necessary optimality conditions along with the transversality condition, a boundary value problem is obtained. Finally, a numerical approach to tackle the boundary value problem is required. Here we propose the Lanczos spectral Tau method to obtain an approximate solution of these problems exploiting the Tau toolbox software library, which allows for ease of use as well as accurate results.

Keywords: Tau method; nonholonomic systems

1. Introduction

Nonholonomic systems are a class of nonlinear systems that cannot be stabilized by a continuous time-invariant feedback, i.e., at a certain time or state there are constraints imposed on the motion (nonholonomic constraints). These systems are controllable but they cannot move instantaneously in certain directions. They belong to a class of nonlinear differential systems with nonintegrable constraints imposed on the motion [1].

Nonholonomic control systems, which result from formulations of nonholonomic systems that include control inputs, are nonlinear control problems requiring nonlinear treatment. There is ample literature on the formulation of the equations of motion and on the dynamics of nonholonomic systems, being [2] an excellent survey for examples. Nonholonomic control systems have been studied in the context of robot manipulation, mobile robots, wheeled vehicles, and space robotics, just to mention a few. In the case of wheeled vehicles, the kinematics and dynamics can be modeled based on the assumption that the wheels are ideally rolling. Typical constraints of wheeled vehicles are rolling contact, like rolling between the wheels and the ground without slipping, or sliding contact such the sliding of skates.

The solution of nonholonomic optimal control problems can be obtained following a standard procedure, which consists of applying Pontryagin's maximum/minimum principle to obtain set of equations along with initial and terminal conditions resulting into a two-point boundary value problem (BVP) [3].

In this work we propose the use of the spectral Tau method to obtain approximate solutions of nonholonomic optimal control problems through the associated BVP.

The spectral Tau method produces a polynomial approximation of the solution of the differential problem. It is based on solving a system of linear algebraic equations, obtained by imposing that all conditions are verified exactly and the residual is orthogonal to the first elements of an orthogonal polynomial basis [4].

The paper is organized as follows. Section 2 describes the system model of the nonholonomic wheeled vehicle and Section 3 explains the optimal control formulation. A brief description of the Tau method is presented in Section 4. An illustrative example with numerical results is provided in Section 5 and some conclusions are drawn in Section 6.

2. Nonholonomic Wheeled Vehicle Model

Vehicle models are usually described by a set of ordinary differential equations that define the dynamics of the vehicle and the relationship between the state variables and control input. The kinematic model of a wheeled vehicle can be defined by the following differential equations

$$\dot{x} = f_1(x, y, v, \theta, \phi)$$
$$\dot{y} = f_2(x, y, v, \theta, \phi)$$
$$\dot{\theta} = f_3(x, y, v, \theta, \phi),$$

where $(x, y) \in \mathbb{R}^2$ is the robot's position in space, θ is the angle with respect to the x-axis, ϕ is the steering wheel's angle with respect to the robot's longitudinal axis, and v is the velocity. (see Figure 1).

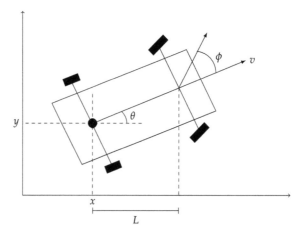

Figure 1. Car-like robot model.

A nonholonomic car-like robot is a car model which rolls without slipping between the wheels and the ground. This constraint is expressed by the equation [5]

$$\dot{x} \sin(\theta) - \dot{y} \cos(\theta) = 0. \tag{1}$$

The simplest model corresponds to a robot with a single wheel: the unicycle model. In this model the wheel rolls on a plane while keeping its body vertical. It is an unrestricted model since it can rotate freely while standing in its position (x, y). Furthermore, the dynamics are characterized by

$$\begin{cases} \dot{x} = v\cos(\theta) \\ \dot{y} = v\sin(\theta) \\ \dot{\theta} = \phi. \end{cases} \tag{2}$$

The kinematic model of a car-like robot has the same state variables as the unicycle model and its dynamic is represented by

$$
\begin{cases}
\dot{x} = v cos(\theta) \\
\dot{y} = v sin(\theta) \\
\dot{\theta} = vu,
\end{cases}
\tag{3}
$$

where $u \in [-\frac{1}{r}, \frac{1}{r}]$ stands for the curvature and r for the turning radius of the robot that corresponds the maximum curvature [6].

3. Optimal Control Problems

An optimal control problem (OCP) can be formulated as

$$
\text{Minimize } J(x(t), u(t)) = \int_{t_0}^{t_f} F(x(t), u(t)) \, dt + G(x(t_f), t_f)
$$

subject to

$$
\begin{aligned}
&\dot{x}(t) = f(x(t), u(t)), \quad t \in [t_0, t_f] \\
&x(t_0) = x_0 \\
&x(t_f) = x_f \\
&x(\cdot) \in X \\
&u(\cdot) \in U \\
&t_f \in [t_0, +\infty[,
\end{aligned}
\tag{4}
$$

where J is the cost function, x is the state vector representing the dynamics, u is the control vector, x_0 is the initial configuration and x_f is the final configuration.

The solution of the OCP can be obtained following a standard procedure, which consists in applying Pontryagin's maximum principle and obtaining the necessary optimality conditions along with the transversality condition resulting into a two-point boundary value problem (BVP).

Pontryagin Maximum Principle

Considering the Hamiltonian function

$$
H(x(t), u(t), \lambda) = F(x(t), u(t)) + \lambda f(x(t), u(t)),
\tag{5}
$$

where F and f are the functions described above and $\lambda = [\lambda_1(t), \lambda_2(t), \dots, \lambda_n(t)]$ is a vector of co-state variables, and considering that (x, u^*) is a controlled trajectory defined over the interval $[t_0, t_f]$ then (x, u^*) is optimal, for all admissible controls u, if the Pontryagin's maximum principle holds, i.e.,

$$
H(x, u^*, \lambda) \geq H(x, u, \lambda).
$$

The Pontryagin maximum principle guarantees that if (x, u^*) is an optimal pair, a solution of the problem (4), then the first order necessary conditions

$$
H_x = -\dot{\lambda}
\tag{6}
$$

together with the stationary conditions

$$
H_u = 0
\tag{7}
$$

satisfies the Hamiltonian maximization with transversality conditions given by [3]

$$\lambda(t_f) = 0 \qquad \text{if} \qquad t_f = \infty \text{ or } G(.) = 0$$

or

$$\lambda(t_f) = \left.\frac{\partial G}{\partial x}\right|_{t=t_f}. \tag{8}$$

This reduces the constrained problem (4) to an unconstrained differential equations system (6)–(8). Usually nonlinear, this system of differential equations can be approximately solved by a numerical method. The next section is devoted to introducing the Tau method, which will be used to numerically tackle the problem.

4. Tau Method

The spectral Tau method produces a polynomial approximation, $y_n(x)$, of the solution, $y(x)$, of a given differential problem $Dy(x) = f(x)$, satisfying a set of conditions defined on an interval $]a, b[$. Introduced by Lanczos in 1938 [7] to compute approximate solutions of linear differential problems with polynomial coefficients and right-hand side, the Tau method solves a tuned system of linear algebraic equations obtained by imposing that the conditions are verified exactly and that the residual is minimized in a quadrature sense, i.e, is orthogonal to the first elements of an orthogonal polynomial basis. It can be applied, indifferently, to initial, boundary or mixed value problems and it can be implemented with any orthogonal basis. We begin by introducing the method for the original case and then shed some light on how to extend it for the solution of nonlinear problems with non-polynomial coefficients.

4.1. Preliminaries and Notation

Let $\mathbb{P}$ be the space of all algebraic polynomials and let $D : \mathbb{P} \to \mathbb{P}$ be a linear differential operator of order $\nu \geq 1$ with polynomial coefficients represented by

$$D \equiv \sum_{r=1}^{\nu} p_r \frac{d^r}{dx^r} \tag{9}$$

and y be the exact solution of the differential problem

$$\begin{cases} Dy(x) = f(x), & a < x < b \\ g_j(y) = \sigma_j, & j = 1, \ldots, \nu, \end{cases} \tag{10}$$

where $f \in \mathbb{P}$ is a polynomial or a convenient polynomial approximation and g_j are ν linear functionals, acting on $C^\nu[a, b]$, representing the (supplementary) conditions.

The main idea of the Tau method is to approximate y by the polynomial y_n, solution of the perturbed problem

$$\begin{cases} Dy_n(x) = f(x) + \tau_n(x), & a < x < b \\ g_j(y_n) = \sigma_j, & j = 1, \ldots, \nu, \end{cases} \tag{11}$$

where τ_n is a polynomial perturbation close to zero in $]a, b[$. Choosing an orthogonal polynomial basis $\mathbf{P} = [P_0, P_1 \ldots]$, then the coefficients of y_n are determined imposing that τ_n is orthogonal to P_i, $i = 0, 1, \ldots, n - \nu$.

The original idea of the Tau method [8] is based on the minimax property of Chebyshev polynomials and on the fact that the solution y_n of (11) depends continuously on the residual τ_n. Later generalized to more general bases [4], the method looks for a residual τ_n that minimizes the weighted norm $||.||_w$ associated to the sequence **P**. Indeed, **P** is orthogonal with respect to the weight function $w(x)$ on $[a, b]$

$$\langle P_i, P_j \rangle = \int_a^b w(x)P_i(x)P_j(x)dx = w_i \delta_{ij},$$

where $w_i = \langle P_i, P_i \rangle = ||P_i||_w^2$ and δ_{ij} is the Kronecker delta, and (11) is achieved by imposing

$$\langle \tau_n, P_j \rangle = 0, \ j = 0, 1, \dots, n - v,$$

that is, $\tau_n = \mathcal{O}(P_{n-v})$.

Using suitable matrices, see for example [4,9,10], the differential problem (11) is translated into an algebraic problem. Such matrices must be computed with criteria in order to ensure stable computations [10].

4.2. Nonlinear Problems

Nonlinear differential problems are solved iteratively by first linearizing the problem and then applying the Tau method to the linear inner problem.

Let

$$F(y) = 0, \ x \in]a, b[\tag{12}$$

be a differential equation, where F is a differential operator that could be nonlinear in y and on its derivatives.

From y_m, an approximation of the exact solution y, we take the first order Taylor polynomial of F centered in y_m to approximate F

$$D_m y = F(y_m) + \sum_{k=0}^{v} (y^{(k)} - y_m^{(k)}) \frac{\partial F}{\partial y^{(k)}} \Big|_{y_m}. \tag{13}$$

F can be replaced by D_m in (12) to solve

$$\sum_{k=0}^{v} y^{(k)} \frac{\partial F}{\partial y^{(k)}} \Big|_{y_m} = -F(y_m) + \sum_{k=0}^{v} y_m^{(k)} \frac{\partial F}{\partial y^{(k)}} \Big|_{y_m}. \tag{14}$$

Applying the Tau method to the linear differential Equation (14), an iterative process is implemented to get increasingly better approximations for the differential problem

$$D_m y_{m+1} = 0, \ m = 0, 1, \dots \tag{15}$$

For additional details on the use of the Tau method for nonlinear problems the reader is invited to read [11].

5. Numerical Experiments

The proposed example is based on the work of [12] and it will be tackled by the Tau method, described in Section 4.

The problem at hand is an optimal control problem of the form (4) with $F = \frac{1}{2}(u_1^2 + u_2^2)$ and $G = \frac{1}{2}\mathbf{x}^T(t_f)Q\mathbf{x}(t_f)$ where **x** is the vector of state variables and Q is the weighting matrix.

5.1. System Model

Consider an automated vehicle that moves on a horizontal plane, the contact of each wheel with the floor is assumed to satisfy the rolling without slipping condition and the control inputs are the torque generated by two motors mounted on the wheels. For a fixed final time, it is desired to find the control inputs that minimizes the energy of the final state. This system can be modeled by

$$J = \min \int_{t_0}^{t_f} \frac{1}{2}(u_1^2 + u_2^2)dt + \frac{1}{2}\mathbf{x}^T(t_f)Q\mathbf{x}(t_f) \tag{16}$$

subject to

$$m\ddot{x} = \frac{\cos(\theta)}{R}(\mu_1 + \mu_2) - \lambda\sin(\theta) \tag{17}$$

$$m\ddot{y} = \frac{\sin(\theta)}{R}(\mu_1 + \mu_2) + \lambda\cos(\theta) \tag{18}$$

$$I\ddot{\theta} = \frac{L}{R}(\mu_1 - \mu_2) \tag{19}$$

with nonholonomic constraint

$$\dot{x}\sin(\theta) - \dot{y}\cos(\theta) = 0,$$

where the position coordinates (x, y), and the heading angle θ, define the system configuration. The mass m, the inertia I, the wheels' radius R, the half-length of the axis L, are parameters of the system and μ_1 and μ_2 are torques generated by the motors.

Defining the control inputs $\mathbf{u} = [u_1\ u_2]^T$ as

$$u_1 = \frac{1}{mR}(\mu_1 + \mu_2), \quad u_2 = \frac{L}{IR}(\mu_1 - \mu_2)$$

and the state variables $\mathbf{x} = [x_1\ x_2\ x_3\ x_4\ x_5]^T$ as

$$x_1 = x\cos(\theta) + y\sin(\theta)$$
$$x_2 = \theta$$
$$x_3 = -x\sin(\theta) + y\cos(\theta)$$
$$x_4 = \dot{x}\cos(\theta) + \dot{y}\sin(\theta) - \dot{\theta}(x\sin(\theta) - y\cos(\theta))$$
$$x_5 = \dot{\theta}.$$

Equations (17)–(19) can be reduced to the following system of differential equations:

$$\begin{cases} \dot{x}_1 = x_4 \\ \dot{x}_2 = x_5 \\ \dot{x}_3 = -x_1 x_5 \\ \dot{x}_4 = -x_1 x_5^2 + u_1 + u_2 x_3 \\ \dot{x}_5 = u_2. \end{cases}$$

Applying the Hamiltonian H, defined in (5),

$$H(\mathbf{x}, \mathbf{u}, \lambda) = \frac{1}{2}(u_1^2 + u_2^2) + \lambda\dot{x}$$

with $\lambda = [\lambda_1 \; \lambda_2 \; \lambda_3 \; \lambda_4 \; \lambda_5]$, and calculating the necessary conditions (6), the stationary conditions (7) and the transversality conditions (8), the following second order system of differential equations is obtained:

$$
\begin{cases}
x_1 \dot{x}_2 + \dot{x}_3 & = 0 \\
\ddot{x}_1 - \dot{x}_2 \dot{x}_3 - \ddot{x}_2 x_3 + \lambda_4 & = 0 \\
\ddot{x}_2 + x_3 \lambda_4 + \lambda_5 & = 0 \\
-\dot{x}_2 \lambda_3 - \dot{x}_2^2 \lambda_4 + \ddot{\lambda}_4 & = 0 \\
\ddot{x}_2 \lambda_4 + \dot{\lambda}_3 & = 0 \\
-x_1 \lambda_3 + 2\dot{x}_3 \lambda_4 + \lambda_2 + \dot{\lambda}_5 & = 0
\end{cases}
$$

for $\dot{x}_1 = x_4$, $\dot{x}_2 = x_5$, $\dot{\lambda}_4 = -\lambda_1$ and $\dot{\lambda}_2 = 0$, with initial and transversality conditions given by

$$
\begin{cases}
x_i(t_0) = x_{i0}, & i \in \{1,2,3\} \\
\dot{x}_1(t_0) = x_{40} \\
\dot{x}_2(t_0) = x_{50} \\
\lambda_i(t_f) = \lambda_{if}, & i \in \{3,4,5\} \\
\dot{\lambda}_4(t_f) = -\lambda_{1f}.
\end{cases}
\tag{20}
$$

Since this is a nonlinear differential system, in order to implement the Tau method, differential equations need to be linearized. Expressions of the form uv and uv^2 will be replaced, respectively, by

$$
uv \approx v_m u + u_m v - u_m v_m
$$
$$
uv^2 \approx v_m^2 u + 2u_m v_m v - 2u_m v_m^2.
$$

Thus, the differential system becomes

$$
\begin{cases}
\dot{x}_{2,m} x_1 + x_{1,m} \dot{x}_2 + \dot{x}_3 & = x_{1,m} \dot{x}_{2,m} \\
\ddot{x}_1 - \dot{x}_{3,m} \dot{x}_2 - x_{3,m} \ddot{x}_2 - \ddot{x}_{2,m} x_3 - \dot{x}_{2,m} \dot{x}_3 + \lambda_4 & = -\dot{x}_{2,m} \dot{x}_{3,m} - \ddot{x}_{2,m} x_{3,m} \\
\ddot{x}_2 + \lambda_{4,m} x_3 + x_{3,m} \lambda_4 + \lambda_5 & = x_{3,m} \lambda_{4,m} \\
(\lambda_{3,m} + 2\dot{x}_{2,m} \lambda_{4,m}) \dot{x}_2 + \dot{x}_{2,m} \lambda_3 + \dot{x}_{2,m}^2 \lambda_4 + \ddot{\lambda}_4 & = \dot{x}_{2,m} \lambda_{3,m} + 2\dot{x}_{2,m}^2 \lambda_{4,m} \\
\lambda_{4,m} \ddot{x}_2 + \dot{\lambda}_3 + \ddot{x}_{2,m} \lambda_4 & = \ddot{x}_{2,m} \lambda_{4,m} \\
\lambda_{3,m} x_1 - 2\lambda_{4,m} \dot{x}_3 + x_{1,m} \lambda_3 - 2\dot{x}_{3,m} \lambda_4 - \dot{\lambda}_5 & = x_{1,m} \lambda_{3,m} - 2\dot{x}_{3,m} \lambda_{4,m} + \lambda_{2,0}
\end{cases}
\tag{21}
$$

where $x_{i,m}$, $i = 1, 2, 3$ and $\lambda_{i,m}$, $i = 3, 4, 5$ are approximations to x_1, x_2, x_3 and λ_3, λ_4, λ_5.

The matrix representation of the differential problem (21), together with the conditions (20), is of the form $Ta = b$, where

$$
T = \begin{bmatrix} C \\ D \end{bmatrix} \quad \text{and} \quad b = \begin{bmatrix} s \\ f \end{bmatrix}
$$

with

$$
C = \begin{bmatrix}
\mathbf{P}(t_0) & 0 & 0 & 0 & 0 & 0 \\
0 & \mathbf{P}(t_0) & 0 & 0 & 0 & 0 \\
0 & 0 & \mathbf{P}(t_0) & 0 & 0 & 0 \\
\mathbf{P}'(t_0) & 0 & 0 & 0 & 0 & 0 \\
0 & \mathbf{P}'(t_0) & 0 & 0 & 0 & 0 \\
0 & 0 & 0 & \mathbf{P}(t_f) & 0 & 0 \\
0 & 0 & 0 & 0 & \mathbf{P}(t_f) & 0 \\
0 & 0 & 0 & 0 & 0 & \mathbf{P}(t_f) \\
0 & 0 & 0 & 0 & \mathbf{P}'(t_f) & 0
\end{bmatrix}
$$

where $\mathbf{P}(t_i) = [P_0(t_i), \ P_1(t_i), \ \ldots]$ represents the action of boundary conditions (20) over the polynomial base elements and $\mathbf{P}'(t_i)$ its derivatives. Matrix D represents the differential operator

$$
D = \begin{bmatrix}
\dot{x}_{2,m}(M) & x_{1,m}(M)N & N & 0 & 0 & 0 \\
N^2 & A_{2,2} & A_{2,3} & 0 & I & 0 \\
0 & N^2 & \lambda_{4,m}(M) & 0 & x_{3,m}(M) & I \\
0 & A_{4,2} & 0 & -\dot{x}_{2,m}(M) & A_{4,5} & 0 \\
0 & \lambda_{4,m}(M)N^2 & 0 & N & \ddot{x}_{2,m}(M) & 0 \\
\lambda_{3,m}(M) & 0 & -2\lambda_{4,m}(M)N & x_{1,m}(M) & -2\dot{x}_{3,m}(M) & N
\end{bmatrix}.
$$

Finally,

$$
s = [x_{10}, x_{20}, x_{30}, x_{40}, x_{50}, \lambda_{3f}, \lambda_{4f}, \lambda_{5f}, -\lambda_{1f}]
$$

$$
f = \begin{bmatrix}
x_{1,m}\dot{x}_{2,m} \\
-\dot{x}_{2,m}\dot{x}_{3,m} - \ddot{x}_{2,m}x_{3,m} \\
x_{3,m}\lambda_{4,m} \\
\dot{x}_{2,m}\lambda_{3,m} + 2\dot{x}_{2,m}^2\lambda_{4,m} \\
x_{1,m}\lambda_{3,m} - 2\dot{x}_{3,m}\lambda_{4,m} + \lambda_{20}
\end{bmatrix},
$$

where

$$
\begin{cases}
A_{2,2} = -\dot{x}_{3,m}(M)N - x_{3,m}(M)N^2 \\
A_{2,3} = -\ddot{x}_{2,m}(M) - \dot{x}_{2,m}(M)N \\
A_{4,2} = (\lambda_{3,m}(M) + 2\dot{x}_{2,m}\lambda_{4,m}(M))N \\
A_{4,5} = -\dot{x}_{2,m}^2(M) + N^2
\end{cases}
$$

and M and N are matrices described in [4] representing, respectively, the multiplication and the differentiation operator.

5.2. Numerical Results

In this section we report the numerical results for the example described in Section 5.1 with initial positions $(x, y) = (10, 3)$ and heading angle $0°$. The time interval is $[t_0, \ f_f] = [0, 5]$ and the system parameters used in the simulation are $m = 10$ kg, $I = 1.2$ kg·m^2, $R = 0.05$ m and $L = 0.1$ m. The weighting matrix is $Q = 10I$, where I stands for the identity matrix.

The simulation results were obtained using the Tau Toolbox with Chebyshev polynomials.

The state trajectories x_1, x_2 and x_3 and the optimal trajectory for the position (x, y) are illustrated in Figures 2 and 3. The trajectories were obtained with 5th order Chebyshev polynomials.

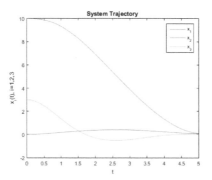

Figure 2. State trajectories x_1, x_2 and x_3.

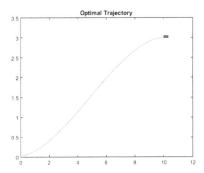

Figure 3. State trajectories x_1, x_2 and x_3.

These approximate solutions only required $m = 8$ iterations to satisfy the stopping criterion $||x_i^m - x_i^{m-1}|| \leq 10^{-14}$. For larger degree polynomials (10 and 15) machine precision can be achieved.

The residual produced by the Tau method in the iteration m is $r_m = Dy_m$, where y_m is the approximating solution of the system of homogeneous differential equations $Dy = 0$. Figure 4 plots the residual $r_{1,m}$, $r_{2,m}$ and $r_{1,m}$ produced by the Tau method for the state variables x_1, x_2 and x_3, respectively.

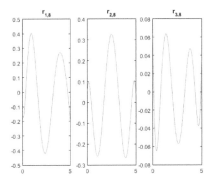

Figure 4. Residual, $r_9 = Dy_9$, produced by the Tau method for the state variables x_1, x_2 and x_3.

Table 1 presents the values for the functional J defined in (16) using polynomials of degree $n = 5, 10$ and 15. Since $u_1 = -\lambda_4$ and $u_2 = \dot{x}_5$, the integral $\int_0^5 \frac{1}{2}(u_1^2 + u_2^2)dt$ and $\mathbf{x}(5)^T Q\mathbf{x}(5)$ can be calculated using the the approximate solutions for the state variables x_i, $i = 1, \ldots, 5$ and the co-state variable λ_4.

As the polynomial multiplication and differentiation, the integration can be set into algebraic operation as well, using a suitable matrix [4].

Table 1. J values for several polynomial degree approximations.

Polynomial Degree	Functional Value
5	5.0990
10	5.0878
15	5.0880

Math. Comput. Appl. **2019**, 24, 91

6. Conclusions

The Lanczos spectral Tau method was used to compute approximate polynomial solutions for nonholonomic systems. A detailed illustration on the approximation procedure is offered. The Tau toolbox provides the appropriate environment to solve systems of ordinary differential problems while allowing for accurate solutions, whenever the sought solution is regular. Numerical results for this dynamical optimization problem confirm both aspects: ease of use and accuracy of approximation.

Author Contributions: A.G. and J.M.A.M. conceived and designed the experiments; J.M.A.M. and P.B.V. performed the experiments; A.G. and J.M.A.M. analyzed the data; A.G. and P.B.V. wrote the paper.

Conflicts of Interest: The authors declare no conflict of interest.

References

1. Fontes, F.A.C.C. Discontinuous feedbacks, discontinuous optimal controls, and continuous-time model predictive control. *Int. J. Robust Nonlinear Control IFAC Affil. J.* **2003**, *13*, 191–209. [CrossRef]
2. Kolmanovsky, I.; McClamroch, N.H. Developments in nonholonomic control problems. *IEEE Control Syst. Mag.* **1995**, *15*, 20–36.
3. Athans, M.; Falb, P.L. *Optimal Control: An Introduction to the Theory and Its Applications*; Courier Corporation: North Chelmsford, MA, USA, 2007.
4. Ortiz, E.L.; Samara, H.J. An operational approach to the Tau method for the numerical solution of non-linear differential equations. *Computing* **1981**, *27*, 15–25. [CrossRef]
5. Angeles, J.; Kecskemethy, A. *Kinematics and Dynamics of Multi-Body Systems*; Springer: Berlin, Germany, 2014.
6. De Luca, A.; Oriolo, G.; Samson, C. Feedback control of a nonholonomic car-like robot. In *Robot Motion Planning and Control*; Springer: Berlin, Germany, 1998; pp. 171–253.
7. Coleman, J. The Lanczos tau-method. *IMA J. Appl. Math.* **1976**, *17*, 85–97. [CrossRef]
8. Lanczos, C. Trigonometric interpolation of empirical and analytical functions. *J. Math. Phys.* **1938**, *17*, 123–199. [CrossRef]
9. Trindade, M.; Matos, J.; Vasconcelos, P.B. Towards a Lanczos'Tau-method toolkit for differential problems. *Math. Comput. Sci.* **2016**, *10*, 313–329. [CrossRef]
10. Vasconcelos, P.; Matos, J.; Trindade, M. Spectral Lanczos' Tau Method for Systems of Nonlinear Integro-Differential Equations. In *Integral Methods in Science and Engineering*; Constanda, C., Dalla Riva, M., Lamberti, P., Musolino, P., Eds.; Birkhäuser: Cham, Switzerland, 2017; Volume 1, pp. 305–314.
11. Gavina, A.; Matos, J.; Vasconcelos, P.B. Improving the Accuracy of Chebyshev Tau Method for Nonlinear Differential Problems. *Math. Comput. Sci.* **2016**, *10*, 279–289. [CrossRef]
12. Yih, C.C.; Ro, P.I. Near-optimal motion planning for nonholonomic systems using multi-point shooting method. In Proceedings of the IEEE International Conference on Robotics and Automation, Minneapolis, MN, USA, 22–28 April 1996; Volume 4, pp. 2943–2948.

Mathematical and Computational Applications

Article

Almost Exact Computation of Eigenvalues in Approximate Differential Problems

José M. A. Matos [1,2,*] **and Maria João Rodrigues** [2,3]

1   Instituto Superior de Engenharia do Porto, Rua Dr. António Bernardino de Almeida, 431,
    4249-015 Porto, Portugal
2   Centro de Matemática da Universidade do Porto, Rua do Campo Alegre, 687, 4169-007 Porto, Portugal;
    mjsrodri@fc.up.pt
3   Faculdade de Ciências da Universidade do Porto, Rua do Campo Alegre, s/n, 4169-007 Porto, Portugal
*   Correspondence: jma@isep.ipp.pt

Received: 29 September 2019; Accepted: 13 November 2019; Published: 14 November 2019

Abstract: Differential eigenvalue problems arise in many fields of Mathematics and Physics, often arriving, as auxiliary problems, when solving partial differential equations. In this work, we present a method for eigenvalues computation following the Tau method philosophy and using Tau Toolbox tools. This Matlab toolbox was recently presented and here we explore its potential use and suitability for this problem. The first step is to translate the eigenvalue differential problem into an algebraic approximated eigenvalues problem. In a second step, making use of symbolic computations, we arrive at the exact polynomial expression of the determinant of the algebraic problem matrix, allowing us to get high accuracy approximations of differential eigenvalues.

Keywords: eigenvalue differential problems; spectral methods; Sturm–Liouville problems

MSC: 34B09; 34L15; 65L15

1. Introduction

Finding eigenfunctions of differential problems can be a hard task, at least for some classical problems. Among others, we can find literature in Sturm–Liouville problems, in Mathieu problems or in Orr–Sommerfeld problems describing the difficulties involved in the resolution of those problems [1–8]. The first difficulty consists of finding accurate numerical approximations for the respective eigenvalues.

In this work, we present a procedure based on the Ortiz and Samara's operational approach to the Tau method described in [9], where the differential problem is translated into an algebraic problem. This is achieved using the called operational matrices that represent the action of differential operators in a function. We have deduced explicit formulae for the elements of these matrices [10,11] obtained by performing operations on the bases of orthogonal polynomials and, for some families, we have exact formulae, which enables the construction of very accurate operational matrices. The Tau method has already been used for these kinds of problems [5,9,12,13]; however, our work on matrix calculation formulas adds efficiency and precision to the method.

Our main purpose is to use the Tau Toolbox, a Matlab numerical library that is being developed by our research group [14–16]. This library allows a stable implementation of the Tau method for the construction of accurate approximate solutions for integro-differential problems. In particular, the construction of the operational matrices is done automatically. These facts led us to think that the Tau Toolbox seems to be useful for these kinds of problems.

Finally, operating with symbolic variables, we define the determinant of those matrices as polynomials and use its roots as eigenvalues' approximations.

We present some examples showing that, using this technique in the Tau Toolbox, we are able to obtain results comparable with those reported in the literature and sometimes even better.

2. The Tau Method

Let $\mathcal{D} : \mathbb{E} \mapsto \mathbb{F}$ be an order ν differential operator, where $\mathbb{E}$ and $\mathbb{F}$ are some function spaces, and let $g_i : \mathbb{E} \mapsto \mathbb{R}$, $i = 1, \ldots, \nu$ be ν functionals representing boundary conditions, so that

$$\begin{cases} \mathcal{D}y = f, & f \in \mathbb{F}, \\ g_i(y) = \phi_i, & i = 1, \ldots, \nu \end{cases} \tag{1}$$

is a well posed differential problem.

2.1. The Tau Method Principle

A particular implementation of the Tau method depends on the choice of an orthogonal basis for $\mathbb{F}$. A sequence of orthogonal polynomials $\{P_n(x)\}_{n=0}^{\infty}$ with respect to the weight function $w(x)$ on a given interval of orthogonality $[a, b]$ satisfies

$$\langle P_i, P_j \rangle = \int_a^b w(x) P_i(x) P_j(x) dx = w_i \delta_{ij},$$

where $w_i = \langle P_i, P_i \rangle$ and δ_{ij} is the Kronecker delta [17].

Let $\mathbb{P}$ be the space of algebraic polynomials of any degree and let us suppose that $\mathbb{P}$ is dense in $\mathbb{F}$; then, the solution y of (1) has a series representation $y \sim \sum_{j \geq 1} a_j P_{j-1}$. A polynomial approximation of degree $n - 1 \in \mathbb{N}$ is achieved by

$$y_n = \sum_{j=1}^{n} a_{n,j} P_{j-1} = \mathcal{P}_n \mathsf{a}_n, \tag{2}$$

where $\mathcal{P}_n = [P_0, P_1, \ldots, P_{n-1}]$ and $\mathsf{a}_n = [a_{n,1}, \ldots, a_{n,n}]^T$.

In the Tau method sense, y_n is a polynomial satisfying the boundary conditions in (1) and solving the differential equation with a residual $\tau_n = \mathcal{D}y_n - f$ of maximal order. Thus, the differential problem is reduced to an algebraic one of finding the n coefficients $a_{n,j}$, $j = 1, \ldots, n$ in (2) such that

$$\begin{cases} g_i(y_n) = \phi_i, & i = 1, \ldots, \nu, \\ \langle P_{i-1}, \mathcal{D}y_n - f \rangle = 0, & i = 1, \ldots, n - \nu, \end{cases} \tag{3}$$

and so the residual $\tau_n = \mathcal{O}(P_{n-\nu})$.

2.2. Operational Formulation

For a given $n \in \mathbb{N}$, $n > \nu$, we define the matrix

$$\mathsf{T}_n = \begin{bmatrix} \mathsf{B}_{\nu \times n} \\ \mathsf{D}_{(n-\nu) \times n} \end{bmatrix} = (t_{i,j})_{i,j=1}^n : t_{i,j} = \begin{cases} g_i(P_{j-1}), \ i = 1, \ldots, \nu \\ \dfrac{\langle P_{i-\nu-1}, \mathcal{D}P_{j-1} \rangle}{w_{i-\nu-1}}, i = \nu+1, \ldots, n \end{cases} \tag{4}$$

and the vector

$$\mathsf{b}_n = \begin{bmatrix} \mathsf{g}_\nu \\ \mathsf{f}_{n-\nu} \end{bmatrix} = (b_i)_{i=1}^n : b_i = \begin{cases} \phi_i, \ i = 1, \ldots, \nu \\ \dfrac{\langle P_{i-\nu-1}, f \rangle}{w_{i-\nu-1}}, i = \nu+1, \ldots, n. \end{cases}$$

If, in problem (1), the differential operator $\mathcal{D}$ is linear and the g_j are ν linear functionals, then problem (3) can be put in matrix form as

$$T_n a_n = b_n.$$

The matrix T_n, called the Tau matrix, can be evaluated from operational matrices, that is, matrices translating into coefficients vectors the action of a differential operator $\mathcal{D}$ in a function y.

Proposition 1. *Let* $\mathcal{P} = [P_0, P_1, \ldots]$ *be an orthogonal polynomial basis,* $y = \mathcal{P}a$ *and* M, N *infinite matrices such that*

$$x\mathcal{P} = \mathcal{P}M \quad and \quad \frac{d}{dx}\mathcal{P} = \mathcal{P}N.$$

Then, for each $k \in \mathbb{N}_0$,

$$x^k y = \mathcal{P}M^k a \quad and \quad \frac{d^k}{dx^k}y = \mathcal{P}N^k a. \tag{5}$$

Proof. For $k = 1$, the result is true by hypothesis. Now, supposing that (5) is true for a $k \in \mathbb{N}$, then

$$x^{k+1}y = x(x^k y) = (x\mathcal{P})M^k a = \mathcal{P}M^{k+1}a$$

and

$$\frac{d^{k+1}}{dx^{k+1}}y = \frac{d}{dx}\left(\frac{d^k}{dx^k}y\right) = \left(\frac{d}{dx}\mathcal{P}\right)N^k a = \mathcal{P}N^{k+1}a,$$

ending the proof by induction. □

The following result generalizes the algebraic representation from the previous proposition to differential operators.

Corollary 1. *Let* $\mathcal{D} : \mathbb{P}_n \mapsto \mathbb{P}_{n+h}$ *be a linear differential operator with polynomial coefficients*

$$\mathcal{D} = \sum_{i=0}^{\nu} p_i \frac{d^i}{dx^i}, \quad p_i \in \mathbb{P}_{n_i} \tag{6}$$

and let $h = \max_{i=0,\ldots,\nu}\{n_i - i\}$.
If $y_n = \mathcal{P}_n a_n$, *then* $\mathcal{D}y_n = \mathcal{P}_{n+h}D_{(n+h)\times n}a_n$ *with*

$$D_{(n+h)\times n} = \sum_{i=0}^{\nu} p_i(M)N^i_{(n-i)\times n'}$$

where $p_i(M) = \sum_{k=0}^{n_i} p_{ik}M^k_{(n+h)\times(n-i)}$ *when* $p_i = \sum_{k=0}^{n_i} p_{ik}x^k$, *and* $A^k_{m\times n}$ *denotes the main* $m \times n$ *block of the matrix* $(A_p)^k$ *with* $p = \max\{m, n\}$.

In [9], the authors discussed the application of this operational formulation of the Tau method to the numerical approximation of eigenvalues defined by differential equations. They proved that, for a differential eigenvalue problem, where in (1)

$$\mathcal{D} = \sum_{r=0}^{t} \lambda^r D_r$$

and λ is a parameter, the zeros of $\det(T_n(\lambda))$ approach the eigenvalues of (1).

2.3. Tau Matrices' Properties

Given that we are dealing with a general orthogonal polynomial basis, instead of particular cases like Chebyshev or Legendre, we can only make assumptions about general properties of Tau matrices T_n. Anyway, we can't expect to have symmetric matrices and, in general, they can be considered sparse but with a low level of sparsity.

Since $\mathcal{P}$ in Proposition 1 is an orthogonal basis, then M is the tridiagonal matrix with the coefficients of its three term recurrence relation. Therefore, for problems with polynomial coefficients, matrices $p_i(M)$ of Corollary 1 are banded matrices, with all non-zero elements between the $\pm n_i$ diagonals.

Matrices N are always strictly upper triangular and so $p_i(M)N^i$ are $n_i - i$ upper Hessenberg matrices. The resulting $\mathcal{D}_{(n-v)\times n}$ block of T_n defined in (4) is a general h upper Hessenberg matrix.

Moreover, one advantage of the Tau method is its ability to deal with boundary conditions, allowing the treatment of any linear combination of values of y and of its derivatives for g_i in (1). Thus, the $v \times n$ block $B_{v\times n}$ in T_n is usually dense, with its entries $g_i(P_j)$, made by linear combinations of orthogonal polynomial values $P_j(x_k)$ and of its derivatives $P_j^{(l)}(x_k)$, in prescribed abscissas x_k.

Assembling those blocks $B_{v\times n}$ and $\mathcal{D}_{(n-v)\times n}$ in T_n, we get an $v + h$ upper Hessenberg matrix.

For some problems, whose dependence on the eigenvalues λ is verified only in the differential equation, we can use Schur complements to reduce matrix sizes. Considering matrix T_n in (4) partitioned as

$$T_n = \begin{bmatrix} B_1 & B_2 \\ D_1 & D_2, \end{bmatrix}$$

where B_1 is $v \times v$ and the other blocks are partitioned accordingly. If B_1 is non-singular, then

$$\det(T_n) = \det(B_1)\det(D_2 - D_1 B_1^{-1} B_2)$$

and the problem is reduced to solve

$$\det(C_n) = 0, \; C_n = D_2 - D_1 B_1^{-1} B_2, \tag{7}$$

reducing to $n - v$ the problem dimension. In the worst case, when B_1 is singular, we have to work with the $n \times n$ matrix T_n.

In the following sections, we illustrate the application of the Tau method to approximate eigenvalues in some classical problems.

3. Problems with Polynomial Coefficients

Sturm–Liouville problems arise from vibration problems in continuum mechanics. The general form of a fourth order Sturm–Liouville equation is

$$(p(x)y''(x))'' - (q(x)y'(x))' + r(x)y(x) = \lambda\mu(x)y(x), \; a < x < b \tag{8}$$

with appropriate initial and boundary conditions, where $a < b \in \mathbb{R}$, p, q, r, and μ are given piecewise continuous functions, with $p(x) > 0$ and $\mu(x) > 0$. These conditions mean that (8) has an infinite sequence of real eigenvalues, bounded from above, and each one has multiplicity of at most 2 [1].

If p and q are differentiable functions, it is an elementary task to give (8) the form

$$p(x)y^{(4)}(x) + 2p'(x)y'''(x) + (p''(x) - q(x))y''(x) - q'(x)y'(x) + (r(x) - \lambda\mu(x))y(x) = 0.$$

From this equation, we derive the operational matrix for the general form of the fourth order Sturm–Liouville differential operator associated with (8)

$$D = p(M)N^4 + 2p'(M)N^3 + (p''(M) - q(M))N^2 - q'(M)N + r(M) - \lambda\mu(M). \tag{9}$$

Assuming that coefficients p, q, r, and μ are polynomials, or convenient polynomial approximations of the coefficient functions, then the height of this differential operator is well defined as

$$h = \max\{\deg(p) - 4, \deg(q) - 2, \deg(r), \deg(\mu)\},$$

where $\deg(.)$ is the polynomial degree. One consequence of Corollary 1 is that to evaluate the block $D_{(n-v)\times n}$ in (4) we have to apply (9) with M and N truncated to its first $n + h$ lines and columns.

The Tau matrix of a fourth order Sturm–Liouville problem is the $n \times n$ matrix $T_n = [B_{4\times n}; D_{(n-4)\times n}]$, where $B_{4\times n}$ is the $4 \times n$ matrix representing boundary conditions and $D_{(n-4)\times n}$ is the first $(n-4) \times n$ main block of D.

Example 1. *Consider the Sturm–Liouville boundary value problem*

$$\begin{cases} y^{(4)}(x) = \lambda y(x), & 0 < x < 1, \\ y(0) = y(1) = y'(0) = y''(1) = 0, \end{cases} \tag{10}$$

whose exact eigenvalues satisfy [1,2]

$$\tanh(\sqrt[4]{\lambda}) - \tan(\sqrt[4]{\lambda}) = 0. \tag{11}$$

In that case $D = N^4 - \lambda I$, where I is the identity matrix, and the boundary conditions can be represented by $B_n = [v_0; v_1; v_0 N; v_1 N^2]$, where $v_0 = [1, -1, \cdots, (-1)^{n-1}]$ and $v_1 = [1, 1, \cdots, 1]$ are length n line vectors with the polynomial base values in the boundary domain.

For each $n > 4$, C_n in (7) is an $n - 4$ square matrix and its determinant an $n - 4$ degree polynomial. We use the Matlab function *roots* to find its zeros and we inspect their accuracy by testing if they satisfy relation (11).

In Figure 1, we present $|\tanh(\sqrt[4]{\lambda_{n,k}}) - \tan(\sqrt[4]{\lambda_{n,k}})|$ for $k = 1, \ldots, 10$ the first 10 eigenvalues approximations obtained with $n = 21, 28, 35$ and $n = 42$, with Chebyshev and Legendre bases shifted to $[0, 1]$.

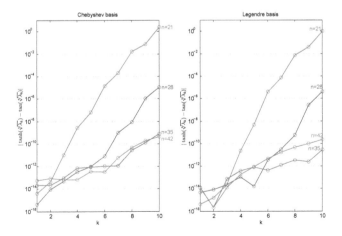

Figure 1. $|\tanh(\sqrt[4]{\lambda_{n,k}}) - \tan(\sqrt[4]{\lambda_{n,k}})|$, $k = 1, 2, \ldots, 10$, $\lambda_{n,k}$ being the roots of $\det(T_n)$ in Example 1.

Example 2. *A very similar problem, presented as the clamped rod problem in [12,18], is*

$$\begin{cases} y^{(4)}(x) = \lambda y(x), & -1 < x < 1, \\ y(\pm 1) = y'(\pm 1) = 0. \end{cases} \tag{12}$$

In that case, and whenever we have a symmetric problem and a symmetric base, the matrix C_n in (7) has zeros intercalating all non-zero elements. We can reduce the problem dimension, defining two matrices C_O and C_E with respectively the odd and the even entries of C_n, then $\det(C_n) = \det(C_O)\det(C_E)$. In that case, since C_n is an 4-upper Hessenberg matrix, C_O and C_E are 2-upper Hessenberg matrices. The sparsity pattern of those two matrices, in Legendre basis and with $n = 52$, are showed in Figure 2.

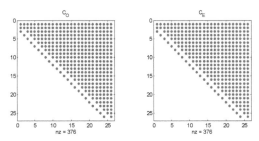

Figure 2. Sparsity pattern of C_O and C_E with $n = 48$ in Legendre base, for Example 2.

The first 14 eigenvalues, evaluated with an 16×16 matrix, are presented in [12]. In Table 1, we compare those values with our results in the Legendre basis and with $n = 16$ and $n = 52$. We present values of $\lambda_{52,k}/k^4$, which allows us to verify that our estimates satisfy the property that the kth eigenvalue is proportional to k^4 [18].

Table 1. Eigenvalues of Example 2 presented in [12] and $\lambda_{n,k}$ with $n = 20$ and $n = 52$ in Legendre basis.

k	λ_k [12]	$\lambda_{16,k}$	$\lambda_{52,k}$	$\lambda_{52,k}/k^4$
1	3.12852439×10^1	$3.128524385877707 \times 10^1$	$3.128524385877221 \times 10^1$	31
2	2.377210675×10^2	$2.377210675311160 \times 10^2$	$2.377210675311198 \times 10^2$	15
3	9.136018866×10^2	$9.136018831954221 \times 10^2$	$9.136018831951466 \times 10^2$	11
4	2.4964874758×10^3	$2.496487437857343 \times 10^3$	$2.496487437856835 \times 10^3$	9.8
5	5.5710074688×10^3	$5.570962978086419 \times 10^3$	$5.570962978573987 \times 10^3$	8.9
6	$1.08631975968 \times 10^4$	$1.086758221396396 \times 10^4$	$1.086758221697812 \times 10^4$	8.4
7	$1.93928004466 \times 10^4$	$1.926303581823010 \times 10^4$	$1.926302825661405 \times 10^4$	8.0
8	$3.05369477203 \times 10^4$	$3.178016789424974 \times 10^4$	$3.178009645408997 \times 10^4$	7.8
9	$6.03075735206 \times 10^4$	$4.960468438481630 \times 10^4$	$4.958769590877672 \times 10^4$	7.6
10	$7.11035649235 \times 10^4$	$7.407231618213559 \times 10^4$	$7.400084934912209 \times 10^4$	7.4
11	$3.597677558196 \times 10^5$	$1.091110089932048 \times 10^5$	$1.064806931408837 \times 10^5$	7.3
12	$3.856105241227 \times 10^5$	$1.548144789089380 \times 10^5$	$1.486344772858071 \times 10^5$	7.2
13	$1.62401642422808 \times 10^7$	$2.619042084314734 \times 10^5$	$2.022155654215451 \times 10^5$	7.1
14	$1.71337968904269 \times 10^7$	$3.728032882888538 \times 10^5$	$2.691234348268295 \times 10^5$	7.0

Example 3. *Consider the following Sturm–Liouville problem with non-null q and r coefficients*

$$\begin{cases} y^{(4)}(x) - (\alpha x^2 y'(x))' + (\beta x^4 - \alpha)y(x) = \lambda y(x), & 0 < x < 5, \\ y(0) = y''(0) = y(5) = y''(5) = 0, \end{cases} \tag{13}$$

with constants $\alpha, \beta \in \mathbb{R}$.

The operational matrix (9) for this case is

$$D = N^4 - \alpha M^2 N^2 - 2\alpha MN + \beta M^4 - (\alpha + \lambda)I$$

and $B_n = [v_0; v_0 N^2; v_1; v_1 N^2]$, with the same v_0 and v_1 vectors of the previous example.

If $\lambda_{n,k}$ is the kth root of $\det(T_n)$, and considering $\tilde{\delta}_{n,k} = \frac{\lambda_{n,k} - \lambda_{n-1,k}}{\lambda_{n,k}}$ as an estimative of the relative error in $\lambda_{n-1,k}$, then $\delta_n = \max_{k=1,\dots,m} |\tilde{\delta}_{n,k}|$ is an estimative of the maximum relative error in the first m eigenvalues of the problem. In Figure 3 left, we present δ_n, with $m = 6$ and with $m = 8$ for Example 3 with $\alpha = 0.02$ and $\beta = 0.0001$ for $n = 16, \dots, 45$. In Figure 3 right, the absolute relative error $|\tilde{\delta}_{n,1}|$ in the lowest eigenvalue is presented for the same n values.

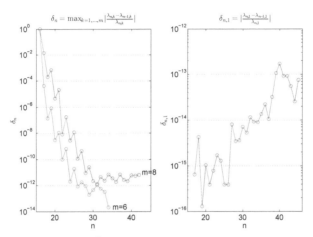

Figure 3. $\delta_n = \max_{k=1,\dots,m} \tilde{\delta}_{n,k}$, $m = 6$ and $m = 8$, (left) and $\delta_{n,1}$ (right), in Example 3.

In Table 2, we compare our results with those of [2] for the first six eigenvalues, and of [8] for the first 4, obtained with values $\alpha = 0.02$ and $\beta = 0.0001$.

Table 2. Eigenvalues of Example 3 presented in [8] and [2] and $\lambda_{n,k}$ with $n = 35$.

k	λ_k [8]	λ_k [2]	$\lambda_{n,k}$
1	0.21505086437	0.21505086436971492	0.2150508643697136
2	2.75480993468	2.754809934682884	2.754809934683077
3	13.2153515406	13.215351540558812	13.21535154055782
4	40.9508193487	40.95081975913755	40.95081975916199
5		99.05347813813880	99.05347806349896
6		204.35449348957833	204.3557322681771

Example 4. *Now, we consider the Orr–Sommerfeld problem*

$$\begin{cases} y^{(4)}(x) - 2\alpha^2 y'' + \alpha^4 y = i\alpha R[(U - \lambda)(y'' - \alpha^2 y) - U'' y], & -1 < x < 1, \\ y(\pm 1) = y'(\pm 1) = 0, \end{cases} \tag{14}$$

with fixed constants α, R and function U.

The particular case $U = 1 - x^2$ is the Poiseuille flow and, with $\alpha = 1$ and $R = 10000$ was treated in [3–5,12]. The operational matrix in that case is

$$D = N^4 - [(2\alpha^2 + (1 - \lambda)i\alpha R)I - i\alpha RM^2]N^2 + (\alpha^4 + (1 - \lambda)i\alpha^3 R - 2i\alpha R)I - i\alpha^3 RM^2.$$

Like in Example 2, this is an upper Hessenberg matrix with zeros intercalating its non-zero elements and we can reduce the problem dimension, splitting in two matrices the Schur complement C_n of the resulting Tau matrix T_n. Choosing Chebyshev basis, this is the operational version of the Tau procedure of [5], where the author was confined to eigenvalues associated with symmetric eigenfunctions, which is equivalent to finding the eigenvalues of C_E.

In [5], the author obtained for $\lambda_1 = 0.23752649 + 0.00373967i$ as an 8 decimal places exact value for the most unstable mode of this problem. Working with double-precision arithmetic, we obtain $\lambda_1 = 0.2375264889204038 + 0.0037396707040170985i$. This value results with $n = 58$ that is an 29×29 matrix C_E, the same dimension used in [5].

In addition, with $R_c = 5772.22$, the smallest value of R for which an unstable eigenmode exists [5], and $\alpha_c = 1.02056$, we get the results presented in Table 3, together with those of [5].

Table 3. Values of λ_1 of Example 4 with critical values $R_c = 5772.22$ and $\alpha = 1.02056$.

n	λ_1
50	$0.2640017404987603 - 3.099641201518763i \times 10^{-9}$
60	$0.2640017396216782 - 3.035288544655118i \times 10^{-9}$
80	$0.2640017390806805 - 2.028898296212456i \times 10^{-9}$

n	λ_1 [5]
44	$0.26400174 - 1.7i \times 10^{-9}$
50	$0.26400174 + 5.9i \times 10^{-10}$

4. Non-Polynomial Coefficients

In the previous section, we solved problems in the conditions of Corollary 1, i.e., with differential operators acting in polynomial spaces. In a more general situation, if some of the coefficients p_i in (6) are non-polynomial functions, then the corresponding matrices $p_i(M)$ are functions of M instead of polynomial expressions.

If a non-polynomial function p_i in (6) can be defined implicitly by a differential problem, with polynomial coefficients, then we can first of all use the Tau method to find a polynomial approximation $\tilde{p}_i$ to p_i and use $\tilde{p}_i(M)$ to approximate the matrix $p_i(M)$.

Example 5. *Mathieu's equation appears related to wave equations in elliptic cylinders [19]. For an arbitrary parameter q, the problem is to find the values of λ for which non-trivial solutions of*

$$y''(x) + (\lambda - 2q\cos(2x))y(x) = 0 \tag{15}$$

exist with prescribed boundary conditions.

It can be shown that there exists a countably infinite set of eigenvalues a_r associated with even periodic eigenfunctions and a countably infinite set of eigenvalues b_r associated with odd periodic eigenfunctions [19]. We are interested in reproducing some of those values given in there.

The operational matrix for problem (15) is

$$D = N^2 + \lambda I - 2q\cos(2M).$$

Our first step to approximate Mathieu's eigenvalues is to approximate matrix $\cos(2M)$. This can be done by, firstly, considering the function $z(x) = \cos(2x)$ as the solution of a differential problem, using Tau method to get a polynomial approximation $z_n(x) \approx z(x)$. In a second step, the operational matrix D is approximated by

$$\tilde{D} = N^2 + \lambda I - 2q z_n(M)$$

and, finally, the last step consists in building the Tau matrix T_n and evaluating the zeros of its determinant.

We take integer values $q = 0, 1, \ldots, 16$ and boundary conditions $y'(-1) = y'(\pi/2) = 0$ to get $a_r(q)$ for even r, $y'(-1) = y(\pi/2) = 0$ for odd r, and $y(-1) = y(\pi/2) = 0$ to get $b_r(q)$ for odd r and $y(-1) = y'(\pi/2) = 0$ for even r.

In Figure 4, we show Mathieu eigenvalues $a_r(q)$, $r = 0, \ldots, 5$ and $b_r(q)$, $r = 1, \ldots, 6$. Values were obtained with a 18th degree polynomial approximation $z_{18} \approx \cos(2x)$ and a 36×36 Tau matrix T_{36} in Chebyshev polynomials.

We can observe, as pointed out in [19], that, for a fixed $q > 0$, we have $a_0 < b_1 < a_1 < b2 < \cdots$ and that $a_r(q)$, $b_r(q)$ approach r^2 as q approaches zero.

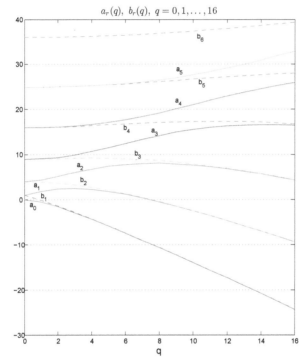

Figure 4. Mathieu eigenvalues $a_r(q)$, $r = 0, \ldots, 5$ and $b_r(q)$, $r = 1, \ldots, 6$, for $q = 0, \ldots, 16$ in Example 5.

Example 6. *Mathieu's equation also appears coupled with a modified Mathieu's equation in systems of differential equations as multi parameter eigenvalues problems. The particular case*

$$\begin{cases} y_1''(x_1) + (\lambda - 2q\cos(2x_1))y_1(x_1) = 0, \ 0 < x_1 < \frac{\pi}{2}, \\ y_1'(0) = y_1'(\frac{\pi}{2}) = 0, \\ y_2''(x_2) - (\lambda - 2q\cosh(2x_2))y_2(x_2) = 0, \ 0 < x_2 < 2, \\ y_2'(0) = y_2(2) = 0 \end{cases} \tag{16}$$

is studied in [6,7] associated with the eigenfrequencies of an elliptic membrane with semi axes $\alpha = \cosh(2)$ and $\beta = \sinh(2)$.

To approximate eigenvalues for this problem, we first have to approximate $\cos(2x)$ and $\cosh(2x)$ by polynomials. Considering, as in the previous example, $z_{16} \approx \cos(2x)$ the 16th degree Tau solution of

$$\begin{cases} z''(x) + 4z(x) = 0, \ 0 < x < \frac{\pi}{2}, \\ z(0) = 1, \ z(\frac{\pi}{2}) = -1, \end{cases} \tag{17}$$

and $w_{16} \approx \cosh(2x)$ as the same degree Tau solution of

$$\begin{cases} w''(x) - 4w(x) = 0, \ 0 < x < 2, \\ w(0) = 1, \ w'(0) = 0, \end{cases} \tag{18}$$

then

$$\tilde{D}_1 = N^2 + \lambda I - 2qz_{16}(M)$$

and

$$\tilde{D}_2 = N^2 - \lambda I + 2qw_{16}(M)$$

are matrices approximating the operational matrices associated with differential equations (16).

For each fixed q, we define matrices Tau T_1 and T_2, representing Mathieu and modified Mathieu equations, respectively. Defining $a_n(q)$ the nth eigenvalue of T_1, in ascending order, and $A_m(q)$ the mth eigenvalue of T_2, in descending order, in [6], it was proved that $a_n(q)$ and $A_m(q)$ are analytical functions of q. Moreover, for each pair (m, n), it was proved the existence and uniqueness of an intersection point of curves $a_n(q)$ and $A_m(q)$. Those intersections identify the eigenmodes of the elliptic membrane.

In Figure 5, we recover, and extend, figures presented in [6] and in [7]. Intersection points of $a_n(q)$, the almost vertical curves, and $A_m(q)$, the oblique curves, are the eigenpairs (a, q) of (16).

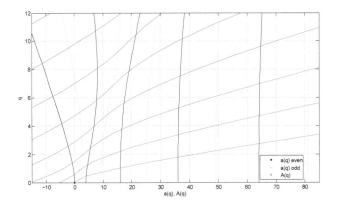

Figure 5. Mathieu eigenvalues $a_n(q)$ and $A_m(q)$, for $q = 0, \ldots, 12$ in Example 5. Only values $a(q), A(q) \in [-15, 85]$ are presented.

5. Nonlinear Eigenvalues Problem

In some differential problems, the eigenvalues can arise in a nonlinear relation with eigenfunctions. Let us consider the following second order problem, related to Weber's equation:

Example 7.

$$\begin{cases} y''(x) + (\lambda + \lambda^2 x^2)y(x) = 0, \ -1 < x < 1, \\ y(-1) = y(1) = 0. \end{cases} \tag{19}$$

Math. Comput. Appl. **2019**, *24*, 96

The operational matrix corresponding to the differential equation is

$$D = N^2 + \lambda I + \lambda^2 M^2$$

and so $\det(T_n)$ is a polynomial with degree $2(n-2)$ in λ.

In Table 4, we present the 10 eigenvalues closest to zero, obtained with $n = 30$ and with $n = 31$. We can verify that $\max_{k=-5,\dots,5} |\lambda_k(30) - \lambda_k(31)| < 5 \times 10^{-9}$.

Table 4. Eigenvalues of Example 7 with $n = 30$ and $n = 31$. Decimal places presented are those that coincide, until to the first distinct two.

k	$\lambda_k(n = 30)$	$\lambda_k(n = 31)$
-5	-19.674904478	-19.674904482
-4	-13.62505355977	-13.62505355969
-3	-13.200062264051	-13.200062264066
-2	-7.0356879747644	-7.0356879747642
-1	-6.59716200235713	-6.59716200235723
1	1.951702364990329	1.951702364990324
2	4.28611106118016	4.28611106118021
3	7.5459203349991	7.5459203349981
4	10.126005915959	10.126005915966
5	13.52870217426	13.52870217408

6. Conclusions

Since the pioneering works of Orzag [5] and Ortiz and Samara [9], the Tau method has been scarcely used to solve differential eigenvalues' problems. With our work, we conclude that the Tau method is a competitive one if we want to evaluate with high accuracy the first eigenvalues, in a large kind of differential problem.

Author Contributions: Formal analysis, J.M.A.M. and M.J.R.; Investigation, J.M.A.M. and M.J.R.; Writing—original draft, J.M.A.M. and M.J.R.; Writing—review and editing, J.M.A.M. and M.J.R.

Funding: This research was partially supported by CMUP (UID/ MAT/ 00144/ 2019), which is funded by FCT (Portugal) with national (MEC) and European structural funds through the programs FEDER, under the partnership agreement PT2020.

Conflicts of Interest: The authors declare no conflict of interest.

References

1. Attili, B.S.; Lesnic, D. An efficient method for computing eigenelements of Sturm–Liouville fourth-order boundary value problems. *Appl. Math. Comput.* **2006**, *182*, 1247–1254. [CrossRef]
2. Farzana, H.; Islam, M.S.; Bhowmik, S.K. Computation of Eigenvalues of the Fourth Order Sturm–Liouville BVP by Galerkin Weighted Residual Method. *J. Adv. Math. Comput. Sci.* **2015**, *9*, 73–85. [CrossRef] [PubMed]
3. Reddy, S.C.; Schmid, P.J.; Henningson, D.S. Pseudospectra of the Orr–Sommerfeld Operator. *SIAM J. Appl. Math.* **1993**, *53*, 15–47. [CrossRef]
4. Pop, I.S.; Gheorghiu, C.I. A Chebyshev–Galerkin Method for Fourth Order Problems. In Proceedings of the International Conference on Approximation and Optimization, Cluj-Napoca, Romania, 29 July–1 August 1996.
5. Orzag, S.A. Accurate solution of the Orr–Sommerfeld stability equation. *J. Fluid Mech.* **1971**, *50*, 689–703. [CrossRef]
6. Neves, A. Eigenmodes and eigenfrequencies of vibrating elliptic membranes: A Klein oscillation theorem and numerical calculations. *Commun. Pure Appl. Anal.* **2009**, *9*, 611–624. [CrossRef]
7. Gheorghiu, C.I.; Hochstenbach, M.E.; Plestenjak, B.; Rommes, J. Spectral collocation solutions to multiparameter Mathieu's system. *Appl. Math. Comput.* **2012**, *218*, 11990–12000. [CrossRef]

8. Chanane, B. Accurate solutions of fourth order Sturm–Liouville problems. *J. Comput. Appl. Math.* **2010**, *234*, 3064–3071. [CrossRef]
9. Ortiz, E.L.; Samara, H. Numerical solution of differential eigenvalue problems with an operational approach to the Tau method. *Computing* **1983**, *31*, 95–103. [CrossRef]
10. Matos, J.M.A.; Rodrigues, M.J.; Matos, J.C. Explicit formulae for derivatives and primitives of orthogonal polynomials. *arXiv* **2017**, arXiv:1703.00743v2.
11. Matos, J.M.A.; Rodrigues, M.J.; Matos, J.C. Explicit Formulae for Intergral-Differential Operational Matrices. submitted for publication, 2019.
12. Gheorghiu, C.I. *Spectral Methods for Differential Problems*; Institute of Numerical Analysis: Cluj-Napoca, Romania, 2007.
13. Charalambides, M.; Waleffe, F. Gegenbauer Tau Methods with and without Spurious Eigenvalues. *SIAM J. Numer. Anal.* **2008**, *47*, 48–68. [CrossRef]
14. Trindade, M.; Matos, J.; Vasconcelos, P.B. Towards a Lanczos Tau-Method Toolkit for Differential Problems. *Math. Comput. Sci.* **2016**, *10*, 313–329. [CrossRef]
15. Trindade, M.; Matos, J.; Vasconcelos, P.B. Dealing with functional coefficients within Tau method. *Math. Comput. Sci.* **2018**, *12*, 183–195. [CrossRef]
16. Vasconcelos, P.B.; Matos, J.; Trindade, M.S. Spectral Lanczos' Tau Method for Systems of Nonlinear Integro-Differential Equations. In *Integral Methods in Science and Engineering, Volume 1: Theoretical Techniques*; Constanda, C., Dalla Riva, M., Lamberti, P.D., Musolino, P., Eds.; Springer: Cham, Switzerland, 2017; pp. 305–314.
17. Chihara, T.S. *An Introduction to Orthogonal Polynomials*; Dover Publications: Mineola, NY, USA, 2011.
18. Funaro, D.; Heinrichs, W. Some Results About the Pseudospectral Approximation of One-Dimensional Fourth-Order Problems. *Numer. Math.* **1990**, *58*, 399–418. [CrossRef]
19. Blanch, G. Mathieu Functions. In *Handbook of Mathematical Functions*; Abramowitz, M., Stegun, I.A., Eds.; Dover Publications: Mineola, NY, USA, 1974; pp. 721–750.

Mathematical and Computational Applications

Article

Factors for Marketing Innovation in Portuguese Firms CIS 2014

Patrícia Monteiro *, Aldina Correia * and Vítor Braga *

School of Management and Technology, Polytechnic of Porto, Rua do Curral, Casa do Curral, Margaride, 4610-156 Felgueiras, Portugal
* Correspondence: 8150194@estg.ipp.pt (P.M.); aic@estg.ipp.pt (A.C.); vbraga@estg.ipp.pt (V.B.)

Received: 28 September 2019; Accepted: 4 November 2019; Published: 22 November 2019

Abstract: Globalization, radical and frequent changes as well as the increasing importance of applying knowledge through the efficient implementation of innovation is critical under the current circumstances. Innovation has been the source of businesses competitive advantage, but it is not restricted to technological innovations, and thus marketing innovation also plays a central role. This is a significant topic in the marketing field and not yet deeply analysed in academic research. The main objective of this study is to understand what factors influence marketing innovation and to establish a business profile of firms that innovate or do not in marketing. We used multivariate statistical techniques, such as, multiple linear regression (with the Marketing Innovation Index as dependent variable) and discriminant analysis where the dependent variable is a dummy variable indicating if the firm innovates or not in marketing. The results suggest that there are several factors explaining marketing innovation, although in this study, we find that the factors contributing the most for marketing innovation are: the Organizational Innovation Index, customer and/or user suggestions, and intellectual property rights and licensing (IPRL). Most of the literature has studied these factors separately. This research studied such factors together, and it is clear that both organizational innovation and IPRL play an important role that drives firms to innovate in marketing, which differs from some literature; customer suggestions help in the process of marketing innovation, as some authors argue that customers do not always know what they want until they have it. In parallel, this study proved to be useful in understanding that the different values for the Marketing Innovation Index display no influence on the results, since they were equivalent when a dummy variable (innovated/not innovated in marketing) was used as a dependent variable. In practice, we realize that the factors are useful to clarify what Portuguese firms innovate or not in marketing, with no different results when we the four marketing innovation levels (design, distribution, advertising and price) are considered.

Keywords: marketing innovation; CIS 2014; multiple linear regression; discriminant analysis

---

1. Introduction

The era of globalization brought radical and frequent changes, as well as a higher recognition of the importance of knowledge through the successful implementation of innovation.

In fact, the changes are constant, and appear in different ways and at an increasing speed. These changes become a challenge for firms which need to, first, identify trends, through well-defined marketing strategies, and subsequently innovate. Innovation, according to the Organization for Economic Cooperation and Development (OECD) and Eurostat [1], requires the implementation of a new or significantly improved product (good or service), or a process, or a new marketing method, or a new organizational method in business practices, within the organization or external relations.

The role of marketing in an organization is very important since it allows increased sales by establishing a long-term relationship with customers. In fact, in addition to financial issues, marketing allows a better understanding of the customer profile leading to co-creation of value.

In order to become more competitive, firms must design new marketing approaches. Marketing innovation is considered by the literature as a non-technological innovation that lacks the same importance as technological innovations (example: product innovation) [2]. According to Mendonça et al. [3] non-technological innovation is an important factor in competitiveness and productivity growth in the economy, specifically in the service industry.

The OECD and Eurostat [1] define marketing innovation as the implementation of a new marketing concept or strategy that differs significantly from existing ones and that has not been previously used.

This study aims to gain a clearer understanding of the role of marketing innovation in Portuguese firms. First, one needs to understand which factors influence and/or impact and secondly to establish a profile of firms regarding marketing innovation.

Marketing innovation is a recent approach with a significant number of publications from 2009 [4]. Therefore, exploring what factors mostly influence Innovation in Marketing is pertinent since the literature contains limited approaches in this regard. According to Correia et al. [5], to achieve the benefits of innovation in terms of economic growth and business competitiveness, it is important to understand its determinants.

Our paper starts with a literature review, that supports the study, followed by the identification of the goals, assumptions and variables used. Subsequently, multivariate analysis of the sample taken from the CIS database (Community Innovation Survey) 2014 was performed and, finally, a connection between the literature and the results of the two statistical techniques: multiple linear regression and discriminant analysis using the SPSS (Statistical Package for the Social Sciences) are assessed.

The results suggest that there are several factors explaining marketing innovation, although, in this study, we find that the factors with higher contribution to marketing innovation are: The Organizational Innovation Index, customer and/or user suggestions and intellectual property rights and licensing (IPRL). In fact, IPRL increase the capacity of marketing innovation in the sense that firms feel more confident in sharing knowledge since they are protected [6]. In turn, the positive contribution of organizational innovation can be explained by the fact that firms increasingly apply improvements in organizational management through innovative marketing measures [7]. Finally, the contribution of customer suggestions and/or users may be related to the fact that they are the consumers of the innovations implemented through products and/or services, so they perceive of what they want to buy [8].

In parallel, this study proved to be useful for understanding that the different values for the Marketing Innovation Index display no influence on our results, since they were equivalent when using a dummy variable (innovated/not innovated in marketing) as dependent variable. In practice, we realize that such factors are useful to classify Portuguese firms that conduct marketing innovation or not, with no different results when one takes the 4 marketing innovation levels (design, distribution, advertising and price) into account.

2. Literature Review and Research Hypothesis

The main objective of this study is to identify the main factors that influence marketing innovation. Therefore, a survey of scientific production was conducted. Firstly, a literature review was carried out aiming to deepen the knowledge about the subject, promoting ideas for research, identifying gaps in the literature and later reviewing it, considering the methodological purpose of this study.

2.1. Marketing, Innovation and Marketing Innovation Concepts

Marketing is one of the most important business areas, in addition to promoting the brand of the firm, accelerating sales and business, it involves customers in the dynamics of the firm allowing a better understanding of the value proposition in a creative way. Modern consumers value the experience the brand can provide through marketing dynamics, in contrast to the price of the product and/or service. As a result, the objective of firms is to establish a lasting relationship, giving importance to the client's opinion and involving them in the business [9].

There are numerous definitions of marketing but one of the most relevant is from the American Marketing Association [10] that defines marketing as "the activity, set of institutions, and processes for creating, communicating, delivering and exchanging offerings that have value for customers, clients, partners, and society in general". In turn, Kotler and Armstrong [11] argue that marketing is a social and management process by which individuals and organizations obtain what they need by creating and exchanging value with each other. In a restricted business context, marketing involves building profitable and valuable trading relationships with customers. Thus, the authors conceptualize marketing as the process by which firms create value and build strong relationships with customers, aiming to return this value to them [11].

Both definitions, regardless of the temporal emergence, point **customers** as focus of the firm and, consequently, marketing practices.

Dantas and Moreira [12] point out that is through innovating that one can design irreverent advertising that captivates **customers**, it allows low-price traps by competitors, namely innovation should be part of the DNA of competitive organizations. They also argue that not to innovate does not mean dying but it means being vulnerable to the most direct competitors, showing the importance of innovation to organizations.

So, what does innovation mean? In a Yesple way, according to the same authors, *"Innovating is creating new things, doing things differently."* The concept of innovation has been approached by several authors and it depends on its application. Table 1 points out some of existing perspectives:

Table 1. Innovation definitions | Source: Own Elaboration.

Definition	Author and Year
"Innovation is defined as the formation of new products or services, new processes, raw materials, new markets and new organizations."	(Schumpeter, 1934) [13]
"Innovation is the specific instrument of entrepreneurship. It is the act that endows resources with a new capacity to create wealth. Innovation, indeed, creates a resource."	(Drucker, 1985) [14]
"Innovation is the embodiment, combination, and/or synthesis of knowledge in novel, relevant, valued new products, processes, or services."	(Leonard and Walter, 1999) [15]
"An innovation is the implementation of a new or significantly improved product (good or service), or process, a new marketing method, or a new organisational method in business practices, workplace organisation or external relations."	(OECD and Eurostat, 2005) [1]
"Innovation is the creation of something that improves the way we live our lives"	(Obama, 2007) [16]
"Innovation is not the result of thinking differently. It is the result of thinking deliberately (in specific ways) about existing problems and unmet needs."	(Razeghi, 2008) [17]

In fact, these definitions are based around 3 main areas: the product (new or improved), processes and organizations (organizational innovation, management or marketing).

The OECD and Eurostat [1] present a structure (Figure 1) that shows innovation as a system and entails the different types of innovation within a firm, the connection of the firm with other organizations and the market demand.

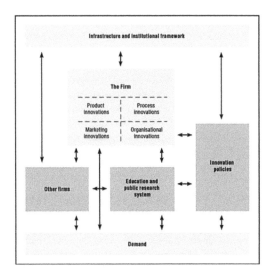

Figure 1. The structure of innovation | Source: [1].

The term innovation has been subject of different adjustments due to its importance in the competitive advantage of firms, thus encompassing fields beyond technological improvements, such as marketing management [18].

In fact, marketing and innovation coexist (Figure 2) and Martin [18] argues that successful modern firms are those that successfully combine innovation and marketing. For example, it is essential to firstly identify trends so that innovation can take place at a subsequent stage, considering what the market and customers need. Indeed, in recent years, new ways of collecting information about consumers through innovative marketing programs have allowed firms to reach their target audience more efficiently by using price strategies that were previously not viable [19].

Figure 2. Marketing innovation | Source: Own Elaboration.

According to Hume et al. [20], Marketing Innovation develops the marketing philosophy throughout the entire innovation process that goes from the emergence of the idea (based on what the customer needs and meets their needs) to the control of the results associated to the launch of the innovation.

On the other hand, the OECD and Eurostat [1] conceptualize marketing innovation as corresponding to the implementation of a new concept or marketing strategy that differs significantly from the existing ones and that has not been previously used by firms. It requires significant changes in appearance/aesthetic or packaging, placement/distribution, promotion or on product pricing policies. It excludes seasonal changes, and regular or other routine changes in marketing methods. This definition is used throughout the present study to support our dependent variable: "Marketing Innovation Index".

2.2. Marketing Innovation and Product Innovation (Good or Service)

According to the OECD and Eurostat [1], product innovation corresponds to the introduction of new goods or services or significantly improved ones in the market, about their abilities or inborn abilities, ease of use, components or subsystems.

Currently, the business community strategically uses different types of innovation; one example is marketing and product innovation. The synergy between both seems to be intuitive, but there are few studies in this area. According to Gupta et al. [21], in their research on the relationship between product innovation and marketing, firms operating product innovation tend to rely on marketing as they face uncertainty about how the product will be understood by consumers. On the other hand, Junge et al. [22] concluded that firms that innovate in the product in parallel with marketing achieve a higher productivity growth. In the same line of thought, Ganzer et al. [23] tried to verify the relationship between skilled labour, turnover and number of employees with the amount of investment in product innovation, innovation process, marketing innovation and organizational innovation and concluded that firms that invest in new products or the improvement of existing products tend to innovate in marketing and, consequently, in the management of the firm. Consequently, our hypothesis is:

Hypothesis 1. *Product innovation contributes positively to marketing innovation.*

Instead, Rebane's [8] study shows different results, since complementarity between product innovation and marketing innovation could not be verified. However, for the services sector the results were different because service providers, when implementing innovation in services and marketing, display greater productivity. Considering these results, the following hypothesis is presented:

Hypothesis 2. *Innovation in services contributes positively to marketing innovation.*

2.3. Marketing Innovation and Organizational Innovation

The OECD and Eurostat [1] show that organizational innovation corresponds to the introduction of a new organizational method in business practices (including knowledge management), in the organization or in the firm's external relations. Higgins [24] mentions that organizational innovation is essential for firms willing to pursue strategic challenges, as they result in improvements in the management of the organization.

The relationship between Organizational Innovation and Marketing Innovation is poorly explored in the literature, but Fleacă et al. [7] studied the extent to which a marketing research process is essential in Organizational Innovation. Their article aimed to understand the importance of using well-defined processes and innovative marketing research, linking the organization's stakeholders to improve work and the overall results of the business.

Marketing research is a sub-process of marketing included in the core processes of a firm, since an effective model of market research allows an organization to more directly and economically commercialize its innovative products, according to current market trends.

The modeling marketing research workflow has drawn valuable results from the APQC (the business process classification framework that allows firms to compare their business processes with other firms [24]). Process classification frameworks developed by the worldwide leader organization in business practice, benchmarking and knowledge management [7].

In this way, a process analyst may be able to structure the necessary steps, such as research objectives, collection, methods and data analysis techniques and information to communicate their findings and implications to those responsible for organizational **decision-making** [7].

Conversely, Ganzer et al. [23] studied the relationship between: product innovation, process, marketing and organization of the knitting industry and concluded that there is a moderate positive correlation between the amount invested in product innovation with the value invested in marketing and organizational innovation. Our hypothesis is:

Hypothesis 3. *Innovative changes in organizational forms contribute to the innovation of marketing techniques.*

2.4. Marketing Innovation and Suggestions of Clients and/or Users in Their Innovation Activities and in the Production of Their Innovative Goods or Services

Clients play a key role in creating and promoting the essential conditions for an innovation project as they allow firms to better understand their needs and desires [25]. Truly, customers are often the consumers of innovations implemented through products and/or services, so they provide important insights about what the market is looking for [8].

Figure 3, proposed by Kilinc et al. [25], reinforces the literature, demonstrating the role of customers in the different stages of the innovation value chain and the impact of the primary roles customers play in the major innovation variables.

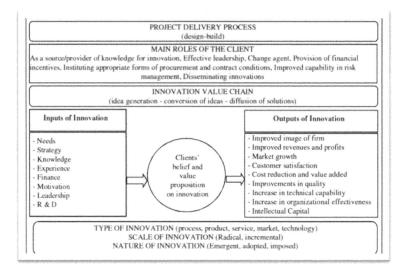

Figure 3. Role of clients | Prepared by: [7].

In contrast, Cabigiosu and Campagnolo [26] report that customers are a source of relevant knowledge, but cannot be used as the main or exclusive source because (i) on one hand, to develop solutions that address the specific needs of customers, there may be a limited match probability of such solutions to other market opportunities and, (ii) on the other hand, according to Tauber [27], customers often do not realize that they need certain innovative products until they are available in the market.

In fact, cooperation with customers may have a positive effect on firms; however, there are still many costs associated with cooperation with customers and negative aspects to introduce radical or revolutionary changes [8].

The literature points to the importance of customer suggestions in innovations and this article aims to understand, in addition to other factors, how customer suggestions contribute to a non-technological innovation [1], such as marketing innovation. Consequently, the following hypothesis is considered:

Hypothesis 4. *Customer suggestions contribute to marketing innovation.*

2.5. Marketing Innovation and Intellectual Property Rights and Licensing

The Oslo manual considers IPRL as requests by firms for patents, European utility models, industrial design rights and trademark registrations [1].

The connection between, for example, registration of brands and product innovation is relatively straightforward and clear, since the marketing of new products is, sometimes, associated with the creation of a new brand to communicate such innovation [3]. As far as marketing innovation is concerned, the connection between them is more complex. According to Mendonça et al. [3] amongst the four types of Innovation in the Oslo manual, only innovation in the promotion of products is not registered, all others can be registered, for example:

* Innovation in aesthetics, appearance and/or packaging: the famous Toblerone packaging is registered for exclusive use;
* Innovation in forms of distribution or sales channels of products: this type of innovation is generally not associated with a brand, except for certain firms, such as Amazon.com;
* Innovation in price: usually this innovation is associated with the telecommunication industry, since price is what distinguishes these types of firms.

Indeed, given the competitiveness of the market, the construction of strong brands may demand marketing innovation, in order to differ from the competition.

In their study, Olaisen and Revang [6] concluded that IPRL increases the innovation capacity, since when IPRL are in place firms feel more confident in sharing knowledge. Also, in this study it was observed that IPRL has no impact on the innovative design of the products. Therefore, the hypothesis for our study is:

Hypothesis 5. *Firms with intellectual property rights and licensing contribute to marketing innovation.*

2.6. Marketing Innovation and Socioeconomic Characteristics of the Firm

The success of innovation can be influenced by the type of organization as well as by the characteristics of its employees [21]. The success of marketing practices depends on the creation of an effective multifunctional team that works as a unit creating value for customers [28]. Consequently, the literature points out that firms involved in product innovation and marketing have qualified employees and with the adequate skills [22,29]. This indication of the literature leads to the hypothesis:

Hypothesis 6. *The academic degree of employees is relevant for marketing innovation.*

Employees are providers of competitive advantage for organizations, and together with turnover they define the size of businesses, i.e., whether the firm is micro, small, medium and/or large. The role of size of the firm is addressed in many studies on Innovation, since it is important to learn about their influence on marketing innovation. Sok et al. [30] state that it is essential, especially, for small and medium enterprises (SMEs) to guarantee the supply of new products, new forms/channels of distribution, to ensure customer satisfaction. The same authors further state that the Yesultaneous implementation of product innovation and marketing combined with qualified employees allow SMEs to be more competitive and achieve better results.

Another aspect leading SMEs to innovate in marketing are circumstantial austerity measures, which do not allow a more permanent support to firms. Therefore, it is imperative that SMEs maximize their internal resources and engage in marketing innovation to better understand the market [31]. In this way the following hypothesis was formulated:

Hypothesis 7. *The business size has an impact on marketing innovation.*

Larger firms are more likely to innovate in marketing techniques than SMEs due to the investment pressure they experience [32]. Notwithstanding the importance of the size of firms, it's also crucial to study the markets in which they operate. The market action defines the strategic path of the firm, so their decisions consider the type of market in which they choose to operate. This factor may contribute to marketing innovation since firms are currently operating in a globalized environment, which forces them, in competitive terms, to modernize and follow the market trends [33]. Thus, we can consider the hypothesis:

Hypothesis 8. *Geographic markets are relevant to firms that innovate in marketing.*

Moreira [33], in his doctoral thesis on the determinants of marketing innovation, conclude that international markets display greater propensity to innovate in Marketing, however, a variable "emerge in national markets" also has a positive and significant effect on innovation marketing. Thus, we can propose the hypothesis:

Hypothesis 9. *Internationalization may explain marketing innovation.*

To achieve a broader explanation for the phenomenon of Marketing Innovation, we will try to understand the synergy between firms that belong to the same innovation group in marketing practices. The literature reports that the effects of synergy between firms of the same group and innovation should be treated with caution due to several factors [34].

However, through the study of Entezarkheir and Moshiri [35] it can be understood that mergers can improve incentives for innovation, promoting economies of scale, increasing the capacity to deal with uncertainty, among other things. It was also concluded that mergers are positively and significantly correlated with firm innovation. Therefore, we try to confirm that:

Hypothesis 10. *Cooperation between firms of the same group is conducive to an innovative marketing environment.*

Figure 4 and Table 2 summarize the research hypothesis, pointed by literature review and considered in this work.

Figure 4. Proposed explanatory model | Source: Own Elaboration.

Table 2. Hypothesis synthesis and theoretical support | Source: Own Elaboration.

Model Variables	Hypothesis	Theoretical Support
Product Innovation (Good and/or Service)	H1—Product innovation contributes positively to Marketing Innovation.	[8,21–23,36]
	H2—Innovation in services contributes positively to marketing innovation.	
Organizational Innovation	H3—Innovative changes in organizational forms contribute to the innovation of marketing techniques.	[7,23]
Customer and/or User Suggestions	H4—Customer suggestions contribute to marketing innovation.	[8,19,25–27]
Intellectual Property Rights and Licensing	H5—Firms that have intellectual property rights and licensing contribute to marketing innovation.	[3,6]
Higher Education of Employees	H6—The formation of the collaborators is relevant for the marketing innovation of a firm.	[22,29]
Business Size	H7—The business size has an impact on marketing innovation.	[30,31]
Geographic Markets	H8—Geographic markets are relevant to firms that innovate in marketing	[33,37,38]
Internationalization	H9—Internationalization is a factor that can help explain the phenomenon of marketing innovation.	
Membership of a Group of Firms (Mergers)	H10—Cooperation between firms of the same group is conducive to an innovative marketing environment.	[34,35]

3. Methodology

The Community Innovation Survey (CIS) 2014 database was used for the study of Marketing Innovation. The CIS is a notation of the National Statistical System regulated by the European Union aiming to measure and characterize innovation activities in European firms. CIS 2014 covers four types of innovation: product innovation, organizational innovation, process innovation and marketing innovation, being this last innovation the focus of this study. This questionnaire is based on Eurostat guidelines and on the principles of the Oslo manual. In fact, this study, in the literature review, tried to approach the definitions contained in the manual whenever possible.

3.1. Population, Sample and Data Collection

The data from CIS 2014 database was the basis of our analysis. Our population was all firms located in Portugal over a period of three years, in which the sample initially consists of 8736 firms and after correction by 7083 valid firms. CIS 2014 collected data on the four types of innovation over the period 2012–2014. The database initially contained 187 variables.

3.2. Exploratory Analysis of Data and Study Variables

Table 3 (Frequency tables and charts in attach) presents a synthesis of the sample used in our study, which was aimed to represent and characterize the data contained in the database. Effectively, it is essential to understand our data before proceeding to multivariate statistical techniques. It can be concluded, from the analysis of Table 3, that most of the Portuguese firms in the sample did not innovate in product, organization and marketing. Within the firms that innovate in marketing, the most frequent innovation is the innovation in the appearance/aesthetic or in the packaging of the products.

Table 3. Exploratory data analysis.

Firms Profile	Classification of economic activity: CAE 46, Wholesale Trade (17.5%) represent a larger share in the sample, followed by CAE 25 Manufacture of Metallic Products (8.7%) and CAE 10 Food Industries (4.5%) (Figure A2).
	Size: considering Decree Law 98/2015, 74.1% corresponds to small firms, 20.7% to medium-sized firms and 5.2% to large firms [39] (Table A1).
	Belongs to a group of firms: 71.7% of the sample, in 2014, did not belong to any group of firms (Table A2).
Geographic Markets	The geographic market is another variable that is of interest for the study, with 16.5% of the sample, between 2012 and 2014, having as a geographic market to sell its goods and/or services the local/regional (MARLOC) market in Portugal, 23.4% the national market (MARNAT) in addition to the regional/local market, 24% market to the European market (MAREUR) and finally 36% to other countries not associated with the European Union (MAROTH) (Table A3).
Higher Education of Employees	Regarding the academic degree of the employees, 25.8% of the firms in the sample have 1 to 4% of the employees with higher education, 20.5% from 10 to 24% and 15.6% do not have any collaborators with higher education (Table A4).
Intellectual Property Rights and Licensing	In the scope of intellectual property and licensing, 85.9% of firms did not require any kind of intellectual property and licensing in the period from 2012 to 2014, from 14.1% requiring 11.2% acquired a patent (PROPAT), 2.2% required a European utility model (PROEUM), 0.5% registered a design right industry (PRODSG)and 0.2% registered a trademark (PROTM) (Table A5)
Marketing Innovation Index	Within the 7 083 valid firms 68.1% did not apply any type of Marketing Innovation, 13.9% applied innovations in the appearance/aesthetics or in the packaging of the products (MKTDGP), 9% in techniques or means of communication for the promotion of goods or services (MKTPDP), 5.1% in the distribution/product placement methods (goods and/or services) or new sales channels (MKTPDL) and 4% in product pricing policies (MKTPRI) (Table A6). Regarding the measures of central trend, the median is 0 meaning that 50% of the firms do not innovate in marketing. The mode is also 0, i.e., the most frequent value, explaining the 68.1% of firms that do not innovate in marketing. The standard deviation is 1.09. The Skewness/Std. Error of Skewness is 59.9, and as it is above 1.96, we conclude that the distribution of the data is asymmetric positive. The Kurtosis/Std. Error of Kurtosis is 34.86 (higher than 1.96), the data distribution is leptokurtic (Table A7).

Table 3. *Cont.*

Organizational Innovation Index	Frequency tables show that 70.5% of the sample did not apply any new organizational method in business practices (including knowledge management), in the organization of the workplace or in the firm's external relations. Of the remaining percentage that applied, 12.7% made new business practices in the organization of procedures (ORGBUP), 9.2% applied new methods of organization of responsibilities and decision-making (ORGWKP) and finally 7.6% innovated in the methods of organization of relations external factors (ORGEXR) (Table A8).
Product and Service Innovation	On the other hand, most of the firms in the sample did not apply any type of innovation in both goods and services (73.5% and 81.5% respectively) (Tables A9 and A10).
Customer and/or User Suggestions 	Most of the sample did not use, during the period between 2012 and 2014, the following means of incorporating the suggestions of customers and/or users: market studies (CLUFEED), consumer groups (CLUMKT), discussion groups and interviews (CLUSUR); surveys of user needs (CLUFOR); development forums (CLUADA); and development of new goods or services by customers and/or users and that the firm has produced and introduced to the market (CLUDEV).
	Most of the sample used the following means of incorporating the suggestions of customers and/or users with the following degrees of importance: customer feedback systems (38.8%); and adaptation of existing goods or services by customers and/or users and the development, production and introduction of these goods or services on the market by the firm with a medium importance level with 27% (Table A11).
Internationalization	Regarding the internationalization of the firms, 58.8% of the sample pointed out that no part of turnover results from sales to customers outside Portugal, 3.6% of firms report that 1% of turnover results from sales to customers outside Portugal and, in turn, 1.7% indicate that 100% of turnover corresponds to sales to customers outside Portugal (Figure A1).

For this study 5 variables were created by the authors in order to investigate the validity of the research hypothesis and to provide the interpretation of the results.

Therefore, the following innovation measures were defined:

* Organizational Innovation Index: this index was calculated from the dummy variables organization of procedures (ORGBUP), organization of responsibilities and decision-making (ORGWKP) and organization of relations external factors (ORGEXR) considering their sum, i.e., Inov_Org = ORGBUP + ORGWKP + ORGEXR, with values between 0 (no item selected) and 3 (all items selected) [5].

* Marketing Innovation Index: this index was calculated from the dummy variables packaging of the products (MKTDGP), promotion of goods or services (MKTPDP), distribution/product placement methods (goods and/or services) or new sales channels (MKTPDL) and product pricing

policies (MKTPRI) considering their sum, i.e., Inov_Mark = MKTDGP + MKTPDP + MKTPDL + MKTPRI, with values between 0 (no item selected) and 4 (all items selected) [5].

Subsequently, the following variables were also created:

* Customer and/or User Suggestions, calculated considering the sum:

Sug_User = market studies (CLUFEED) + consumer groups (CLUMKT) + discussion groups and interviews (CLUSUR) + surveys of user needs (CLUFOR) + development forums (CLUADA) + development of new goods or services by customers and/or users and that the firm has produced and introduced to the market (CLUDEV)

Intellectual Property Rights and Licensing calculated considering the sum: Prop_Intellectual = acquired a patent (PROPAT) + required a European utility model (PROEUM) + registered a design right industry (PRODSG) + registered a trademark (PROTM), with values between 0 (no item selected) and 4 (all items selected).

* Geographic Markets: M_GEO = geographic market to sell its goods and/or services the local/regional market (MARLOC) + national market (MARNAT) + market to the European market (MAREUR) + market to other countries not associated with the European Union (MAROTH), with values between 0 (no item selected) and 4 (all items selected).

3.3. Explanatory Variables

Considering the data analysis and the literature review, a database was built with the variables that could allow a better understanding of Marketing Innovation. Thus, the independent variables pointed out for this multivariate study are summarized in Table 4:

Table 4. Explanatory Variables, Expected Signal and Theoretical Support | Source: Own Elaboration.

Explanatory Variables	Hypothesis	Acronyms	Expected Sign	Theoretical Support
Product Innovation	H1	INPDGD	+	[8,21–23,36]
	H2	INPDSV	+	
Organizational Innovation	H3	Inov_Org	+	[7,23]
Customer and/or User Suggestions	H4	Sug_Users	+	[8,19,25–27]
Intellectual Property Rights and Licensing	H5	Prop_Intellectual	+	[3,6]
Higher Education of Employees	H6	EMPUD	+	[22,29]
Business Size	H7	SIZE14_COD	+	[30,31]
Geographic Markets	H8	M_GEO	+	[33,37,38]
Internationalization	H9	SLO14	+	
Membership of a Group of Firms	H10	GP	+	[34,35]

4. Factors that Influence Marketing Innovation

Multiple linear regression was used for predicting the value of a variable based on the value of two or more variables [40]. The dependent variable was "Marketing Innovation Index". The variables used to predict the value of the dependent variable are the independent variables: GP—"Belonging to a Group of Firms", Inov_Org—"Organizational Innovation Index", M_GEO—"Geographic Markets", Prop_Intellectual—"Intellectual Property Rights and Licensing", Sug_Users—"Customer and/or User Suggestions", INPDGD—"Goods Innovation", INPDSV—"Service

Innovation", EMPUD—"% Of Employees with Higher Education", SLO14—"Internationalization" and SIZE14_COD—"Business Size".

Firstly, we used the forward method in which variables are introduced one by one. The first variable to be introduced is the one with the highest correlation coefficient with the dependent variable Marketing Innovation Index. Subsequently, the variables with the highest coefficient of partial correlation are introduced sequentially [41]. Once the forward analysis was performed it was concluded that the EMPUD, SIZE14_COD, M_GEO and GP variables at a significance level of 5% are not significant for the model (Appendix A Table A12). Consequently, hypothesis H6, H7, H8 and H10 are rejected, i.e., the academic level of employees, the business size, the geographic markets and the probability of belonging to a group of firms do not contribute to explain the Index of Marketing Innovation.

After this, linear regression by the stepwise method was conducted in order to eliminate these variables from the model. By the Stepwise method of the 10 independent variables initially considered, only 6 variables were used for the estimation of the model, and the EMPUD, SIZE14_COD, M_GEO and GP variables were eliminated as expected (Appendix A Table A13).

Analyzing the summary of the multiple linear regression model (Table 5) we conclude that $Ra^2 = 0.204$ so, approximately 20.4% of the Marketing Innovation Index is explained by the independent variables.

Table 5. Summary | linear regression.

Model Summary				
Model	R	R Square	Adjusted R Square	Std. Error of the Estimate
6	0.453 [f]	0.205	0.204	1.10785

[f] Predictors: (Constant), Inov_Org, Sug_Users, Prop_Intellectual, INPDSV, INPDGD, SLO14.

According to the analysis of the ANOVA test (Table 6), *p*-value ≈ 0.000 so, H0 is reject, then we are faced with a highly significant model in which at least one independent variable has a considerable effect on the variation of the dependent variable of marketing innovation.

Table 6. Analysis of variance (ANOVA) test | linear regression.

	ANOVA					
	Model	Sum of Squares	df	Mean Square	F	Sig.
	Regression	1146.055	6	191.009	155.631	0.000 [g]
6	Residual	4445.368	3622	1.227		
	Total	5591.423	3628			

[g] Predictors: (Constant), Inov_Org, Sug_Users, Prop_Intellectual, INPDSV, INPDGD, SLO14.

The variables Organizational Innovation Index (with a standardized coefficient of 0.258), customer suggestions and/or users (with a standardized coefficient of 0.173) and intellectual property and licensing (with a standardized coefficient of 0.147) **are those that contribute** more to explain the Index of Marketing Innovation (Table 7).

Table 7. Model coefficients | linear regression.

Model		Unstandardized Coefficients		Standardized Coefficients	t	Sig.
		B	**Std. Error**	**Beta**		
	(Constant)	0.314	0.035		8.996	0.000
	Inov_Org	0.300	0.018	0.258	16.572	0.000
	Sug_Users	0.046	0.004	0.173	10.947	0.000
6	Prop_Intellectual	0.328	0.034	0.147	9.675	0.000
	INPDSV	0.236	0.043	0.088	5.539	0.000
	INPDGD	0.229	0.040	0.091	5.763	0.000
	SLO14	−0.301	0.066	−0.070	−4.567	0.000

[a] Dependent Variable: Inov_Mark.

> Then, the adjusted model is: Inov_Mark = 0.314 + 0.258 Inov_Org + 0.173 Sug_Users + 0.147 Prop_Intellectual + 0.088 INPDSV + 0.091 INPDGD − 0.070 SLO14

These results are, to some extent, contradictory to the literature review insofar a positive sign was expected for all independent variables (Table 4).

Contrary to expectations (Table 8—NS represent non-significant in the regression model) hypothesis H1 and H10 are rejected, then there is no statistical evidence to consider Product and Organizational Innovation as a factor to Marketing Innovation, as well as H6, H7, H8 and H10.

Table 8. Explanatory Variables, Obtained Signal and Theoretical Support | Source: Own Elaboration.

Explanatory Variables	Hypothesis	Acronyms	Obtained Sign	Theoretical Support
Product Innovation	H1	INPDGD	NS	[8,21–23,36]
	H2	INPDSV	NS	
Organizational Innovation	H3	Inov_Org	+	[7,23]
Customer and/or User Suggestions	H4	Sug_Users	+	[8,19,25–27]
Intellectual Property Rights and Licensing	H5	Prop_Intellectual	+	[3,6]
Higher Education of Employees	H6	EMPUD	NS	[22,29]
Business Size	H7	SIZE14_COD	NS	[30,31]
Geographic Markets	H8	M_GEO	NS	[33,37,38]
Internationalization	H9	SLO14	-	
Membership of a Group of Firms	H10	GP	NS	[34,35]

As expected, organizational innovation, customer and/or user suggestions and intellectual property rights and licensing are proved to be important for increasing marketing innovation, as pointed by literature, as well as internationalization, but the latter with opposite sign to the expected. Thus, taking into account our data, the factors promoting marketing innovation are organizational innovation, customer and/or user suggestions and intellectual property rights and licensing, and internationalization are an obstacle to innovate in marketing.

4.1. Testing the Assumptions of Multiple Linear Regression Analysis

In order to validate the assumptions of the Multiple Linear Regression model, a residual analysis was developed. We analyzed if the residuals follow a normal distribution and had a constant variance (using KS test and dispersion diagrams) and to understand if the residuals are independent, we used the Durbin–Watson test.

Table 9 shows the summary of the multiple linear regression model and the overall adjustment statistics. The Durbin–Watson returned a value of d = 2.002, (approximate to 2) and thus the residuals are not correlated [42]. Consequently, one could proceed with multiple linear regression.

Table 9. Durbin–Watson test | linear regression.

Model Summary [g]					
Model	R	R Square	Adjusted R Square	Std. Error of the Estimate	Durbin-Watson
6	0.453 [f]	0.205	0.204	1.10785	2.002

[f] Predictors: (Constant), Inov_Org, Sug_Users, Prop_Intellectual, INPDSV, INPDGD, SLO14. [g] Dependent Variable: Inov_Mark.

The standard predicted and residual values show approximate maximum and minimum values but are not proportional (Table 10).

Table 10. Residuals statistics | linear regression.

Residuals Statistics [a]					
	Minimum	Maximum	Mean	Std. Deviation	N
Predicted Value	0.0130	3.5917	1.0825	0.57608	4164
Residual	−2.60513	3.45683	−0.01193	1.10424	4164
Std. Predicted Value	−1.869	4.498	0.034	1.025	4164
Std. Residual	−2.352	3.120	−0.011	0.997	4164

[a] Dependent Variable: Inov_Mark.

Through the normal P-P plot of the regression standardized residual in Figure 5, one can conclude that some points are distant from the diagonal. This may indicate that the residuals do not follow a normal distribution.

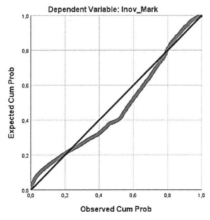

Figure 5. Normal P-P plot of regression standardized residual | linear regression.

In turn, the Scatterplot (Figure 6) presents horizontal lines due to the rounding errors of the values predicted by the regression model for the values of a discrete variable [28].

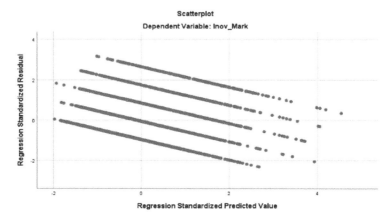

Figure 6. Scatterplot | linear regression.

There is an absence of correlation between independent variables (absence of multicollinearity).

Another assumption for linear regression is that none or few collinearities are present. Collinearity occurs when two independent variables are highly correlated [43].

Table 11 shows that no independent variable presents multicollinearity problems since the T is not adjacent to 0 and the Variance Inflation Factor (VIF) displays values below 5.

Table 11. Collinearity statistics | linear regression.

	Coefficients [a]		
	Model	**Collinearity Statistics**	
		Tolerance	VIF
	(Constant)		
	Inov_Org	0.907	1.102
	Sug_Users	0.881	1.136
6	Prop_Intellectual	0.954	1.048
	INPDSV	0.867	1.153
	INPDGD	0.875	1.143
	SLO14	0.928	1.078

[a] Dependent variable: Inov_Mark.

In the diagnosis of collinearity (Table 12), it follows that the values of the condition index are not close to 30 and the values themselves are distant from 0.

Table 12. Collinearity diagnosis | linear regression.

				Collinearity Diagnostics [a]						
Model	Dimension	Eigenvalue	Condition Index	Variance Proportions						
				(Constant)	Inov_Org	Sug_Users	Prop_Intellectual	INPDSV	INPDGD	SLO14
	1	3.890	1.000	0.01	0.02	0.02	0.02	0.02	0.02	0.02
	2	0.875	2.108	0.00	0.04	0.00	0.18	0.20	0.00	0.35
	3	0.734	2.302	0.01	0.00	0.00	0.78	0.01	0.00	0.24
6	4	0.538	2.688	0.00	0.47	0.02	0.00	0.22	0.27	0.00
	5	0.410	3.082	0.02	0.01	0.05	0.02	0.56	0.37	0.35
	6	0.350	3.332	0.09	0.46	0.34	0.00	0.00	0.31	0.01
	7	0.202	4.392	0.87	0.01	0.57	0.00	0.00	0.02	0.03

[a] Dependent Variable: Inov_Mark.

Most of the proportions of variance, except for a few, show values that are distant from 50%, and may not indicate a multicollinearity problem.

Thus, generically the model meets the multiple linear regression model assumptions, and our model is significant, and there is statistical evidence in the data to consider the conclusions valid.

4.2. Features that Distinguish Firms that Innovate in Marketing

According to Maroco [44], discriminant analysis is "a dependent multivariate technique used to investigate, evaluate differences between groups and classify entities within groups, based on known discretionary variables." In fact, it is used to discriminate between groups, using a categorical dependent variable and independent interval scale variables [45].

As the discriminant analysis aims to discover the characteristics that distinguish the members of one group from members of a different one, the characteristics of a new individual allows predicting the group it belongs to [45]. We aimed to study which are the characteristics of firms that do not innovate in marketing and those that innovate in marketing. In particular we are interested in comparing the results with the previous analysis, where we consider marketing innovation as an index ranging between 0 (no item selected) and 3 (all items selected). For the analysis we considered Marketing Innovation as a dummy variable being 0 for non-innovative in marketing firms and 1 for innovative in marketing firms.

The non-metric dependent variable marketing innovation consists of 2 mutually exclusive categories. The independent metric variables were selected taking the literature into account. Continuously, Figure 7 presents the metric and non-metric variables under study.

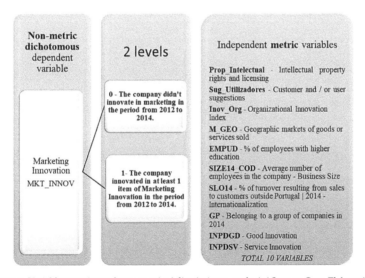

Figure 7. Variables metrics and non-metrics | discriminant analysis | Source: Own Elaboration.

The discriminant analysis requires the verification of the following assumptions:

1. Multivariate normality;
2. Multivariate homoscedasticity;
3. Absence of multicollinearity.

Considering the assumptions, the following tests were performed in order to understand if the discriminant analysis could be performed.

4.2.1. Multivariate Normality

In relation to the first assumption, a K-S test was previously developed and H0 rejected, indicating that the variables do not follow a normal distribution. In order to overcome this problem, we used the central limit theorem which indicates that the larger the size of a sample, the distribution of the mean will be closer to a normal distribution. In this case, the sample contains more than 30 cases so the distribution of the mean can be satisfactorily approximated by a normal distribution [44,46]. The remaining assumptions will then be verified in the output of the discriminant analysis.

4.2.2. Analysis of Variance (ANOVA): Analysis of Differences between Groups

Hypothesis to be tested:

H0: The group averages are equal
H1: The group averages are different

Looking at the test of equality of the groups means, it can be concluded that the Wilks' λ is generally approximate to 1 suggesting that the groups means are equal (Table 13).

Table 13. Tests of equality of group means | discriminant analysis.

	Tests of Equality of Group Means				
	Wilks' Lambda	F	df1	df2	Sig.
GP	1.000	0.221	1	3627	0.638
INPDGD	0.985	54.081	1	3627	0.000
INPDSV	0.980	73.487	1	3627	0.000
SLO14	0.998	5.547	1	3627	0.019
SIZE14_COD	0.999	1.909	1	3627	0.167
EMPUD	0.994	20.648	1	3627	0.000
M_GEO	0.995	17.135	1	3627	0.000
Inov_Org	0.941	228.403	1	3627	0.000
Prop_Intellectual	0.967	122.852	1	3627	0.000
Sug_Users	0.953	177.677	1	3627	0.000

Concerning the F-test, a small value indicates that when independent variables are considered individually, they do not differ between groups. In turn, the variable Inov_Org presents a high F suggesting being a variable that is able to differentiate the groups.

For the significance levels, most variables display a p-value < 0.05, thus rejecting the null hypothesis, i.e., the means in the two groups, of innovative and non-innovative firms, for all variables are equal.

In contrast with the others, the GP and SIZE14_COD present p-values above 5% (Table 13), indicating that these variables probably do not contribute to the model since the null hypothesis cannot be rejected.

4.2.3. Multivariate Homoskedasticity—Box's M test

Hypothesis to be tested:

H0: Equivalent matrices of variance–covariance for the two groups
H1: Different matrices of variance–covariance for the two groups

Analyzing the Box's M test (Table 14), it is verified that the p-value $\approx 0.000 < 0.05$ then rejecting H0, i.e., the variance-covariance matrices are the same for the two groups. Therefore, instead of presenting homoscedasticity, required by the analysis, the data shows heteroscedasticity, becoming a limitation of the analysis.

Table 14. Box's M test | discriminant analysis.

Test Results	
Box's M	576.631
F — Approx.	20.552
F — df1	28
F — df2	43125012.597
F — Sig.	0.000

Tests null hypothesis of equal population covariance matrices.

4.2.4. Absence of Multicollinearity

One of the assumptions of the discriminant analysis is that there is no multicollinearity. Table 15 shows no multicollinearity, i.e., there is no high correlation between the variables, since the values are smaller than 50% presenting, in this case, levels of correlation between variables generally weak [47].

Table 15. Pooled within-groups matrices | discriminant analysis.

		GP	INPDGD	INPDSV	SLO14	SIZE14_COD	EMPUD	M_GEO	Inov_Org	Prop_Intellectual	Sug_Users
						Pooled Within-Groups Matrices					
Correlation	GP	1.000	0.031	0.070	0.045	0.372	0.291	−0.011	0.097	0.013	0.058
	INPDGD	0.031	1.000	0.209	0.185	0.082	0.007	0.209	0.051	0.144	0.174
	INPDSV	0.070	0.209	1.000	−0.127	0.025	0.168	−0.007	0.183	0.012	0.184
	SLO14	0.045	0.185	−0.127	1.000	0.266	−0.118	0.257	0.004	0.118	0.076
	SIZE14_COD	0.372	0.082	0.025	0.266	1.000	0.075	0.086	0.071	0.075	0.100
	EMPUD	0.291	0.007	0.168	−0.118	0.075	1.000	0.062	0.165	0.126	0.125
	M_GEO	−0.011	0.209	−0.007	0.257	0.086	0.062	1.000	0.016	0.150	0.130
	Inov_Org	0.097	0.051	0.183	0.004	0.071	0.165	0.016	1.000	0.048	0.211
	Prop_Intellectual	0.013	0.144	0.012	0.118	0.075	0.126	0.150	0.048	1.000	0.093
	Sug_Users	0.058	0.174	0.184	0.076	0.100	0.125	0.130	0.211	0.093	1.000

4.2.5. Stepwise Method

Since, previously, it was verified that the variables GP and SIZE14_COD do not show significant discriminant power, we used the Stepwise method. This method selects the variables with discriminative capacity, so that the analysis is only done with such variables. In fact, the stepwise method starts without variables and in the following steps variables are added or removed, depending on their discriminative ability [44].

In this analysis the method used for the inclusion/removal of variables was Wilk's λ. Consequently, the variables by this method are included (or removed) according to their inclusion, it greatly decreases (or not) the lambda value [44].

By the stepwise method, only 7 (out of 10) independent variables were considered for the model estimation, and the variables GP, EMPUD and M_GEO were eliminated (Table 16).

Table 16. Variables not considered in the analysis | discriminant analysis.

		Variables Not in the Analysis			
	Step	**Tolerance**	**Min. Tolerance**	**F to Enter**	**Wilks' Lambda**
	GP	0.852	0.794	1.660	0.882
7	EMPUD	0.912	0.850	0.352	0.882
	M_GEO	0.890	0.831	2.875	0.881

Table 17 shows that as variables were introduced, the Wilks' λ decreased. Considering that a variable with little tolerance contributes little to the model, Internationalization (SLO14), shows the smallest tolerance (0.866). Prop_Intellectual and Inov_Org are two variables that present high tolerance values which indicates that they are the ones that most contribute to the model. However, all variables present high tolerance values, thus showing their relevance to the model and the absence of multicollinearity as they approach 1.

Table 17. Variables in the analysis | discriminant analysis.

	Step	Tolerance	F to Remove	Wilks' Lambda
	Variables in the Analysis			
	Inov_Org	0.931	125.117	0.913
	Sug_Users	0.904	68.185	0.899
	Prop_Intellectual	0.964	81.928	0.902
7	INPDSV	0.882	8.671	0.884
	SIZE14_COD	0.917	8.928	0.884
	INPDGD	0.885	12.498	0.885
	SLO14	0.866	10.370	0.885

Hypothesis to be tested:

H0: The group averages are equal

H1: The group averages are different

To understand if the functions are discriminant the Wilks' λ test (Table 18) was performed and it was concluded that one must to reject H0, since the test shows a *p*-value below 5%, i.e., the means of the groups in the function are not equal. Therefore, the functions are discriminant.

Table 18. Wilks' λ | discriminant analysis.

Step	Number of Variables	Lambda	df1	df2	df3	Statistic	df1	df2	Sig.
						Exact F			
7	7	0.882	7	1	3627	69.133	7	3621.000	0.000

To estimate the coefficients of the discriminant function, assuring the significance of the functions

Table 19 shows that there is 1 discriminant function and the eigenvalue attributed to function 1 is 0.134 and represents 100% of the explained variance.

Table 19. Eigenvalues | discriminant analysis.

Function	Eigenvalue	% of Variance	Cumulative %	Canonical Correlation
	Eigenvalues			
1	0.134 [a]	100.0	100.0	0.343

[a] First 1 canonical discriminant functions were used in the analysis.

Regarding canonical correlation, function 1 presents a canonical correlation $(0.343)^2$ corresponding to 0.117649 so, approximately 11.8% of the variance of the groups is explained by the discriminant function 1.

Find the contribution of the variables to the function

Table 20 allows us to understand which variables contribute to the discriminant function. This indicates that for function 1 the variables that most contribute to distinguish innovative from non-innovative firms are Inov_Org, Prop_Intellectual and Sug_Users. On a different perspective, Organizational Innovation Index, intellectual property and licensing and suggestions of clients and/or users display a positive contribution to be classified in the group of firms that innovate in marketing.

Table 20. Standardized canonical discriminant functional coefficients | discriminant analysis.

Standardized Canonical Discriminant Function Coefficients	
	Function
	1
INPDGD	0.182
INPDSV	0.152
SLO14	−0.167
SIZE14_COD	−0.151
Inov_Org	0.552
Prop_Intellectual	0.441
Sug_Users	0.416

In turn, the structured matrix (Table 21) allows examining the contribution (ordered by the absolute value) of each variable to the discriminant function, without the effect of collinearity. In this way, organizational innovation is the factor that most positively contributes to function 1, followed by the intellectual property rights and licensing and customer and/or user suggestions. The results, without the effect of collinearity, remained almost equal to Table 20 since the collinearity test resulted negative for the discriminant analysis.

Table 21. Structured matrix | discriminant analysis.

Structure Matrix	
	Function
	1
Inov_Org	0.686
Sug_Users	0.605
Prop_Intellectual	0.503
INPDSV	0.389
INPDGD	0.334
EMPUD [a]	0.234
M_GEO [a]	0.111
SLO14	−0.107
SIZE14_COD	−0.063
GP [a]	0.036

[a] This variable not used in the analysis.

Classify cases

Table 22 allows observing coefficients by Fisher function which, in turn, allow the classification of cases into groups. Thus, it follows that the classification models:

D0 (Don't Innovate in Marketing) = −3.858 + 0.839 * INPDGD + 0.439 * INPDSV − 0.203 * SLO14 + 3.488 SIZE14_COD + 0.237 * Inov_Org − 0.096 * Prop_Intellectual + 0.155 * Sug_U

D1 (Innovate in at least 1 item of Marketing Innovation) = −4.435 + 1.109 * INPDGD + 0.681 * INPDSV − 0.627 * SLO14 + 3.309 SIZE14_COD + 0.629 * Inov_Org + 0.496 * Prop_Intellectual + 0.222 * Sug_U

Table 22. Classification function coefficients.

Classification Function Coefficients		
	MKT_INNOV	
	No	Yes
INPDGD	0.839	1.109
INPDSV	0.439	0.681
SLO14	−0.203	−0.627
SIZE14_COD	3.488	3.309
Inov_Org	0.237	0.629
Prop_Intellectual	−0.096	0.496
Sug_Users	0.155	0.222
(Constant)	−3.858	−4.435
Fisher's linear discriminant functions		

Interpretation of the results of discrimination and validation

Considering Table 23, 63.7% of the cases were correctly classified. In cross-validation, the percentage is almost the same (63.5%) of the original classification.

Table 23. Classification of results | stepwise discriminant analysis.

Classification Results [a,c]					
		MKT_INNOV			Total
			No	Yes	
Original	Count	No	1010	647	1657
		Yes	669	1303	1972
	%	No	61.0	39.0	100.0
		Yes	33.9	66.1	100.0
Cross-validated [b]	Count	No	1006	651	1657
		Yes	672	1300	1972
	%	No	60.7	39.3	100.0
		Yes	34.1	65.9	100.0

[a] 63.7% of original grouped cases correctly classified; [b] Cross validation is done only for those cases in the analysis. In cross validation, each case is classified by the functions derived from all cases other than that case; [c] 63.5% of cross-validated grouped cases correctly classified.

In Table 24 a comparison between the linear regression model (MLR) and discriminant analysis (DA) results is presented.

As in MLR, organizational innovation, customer and/or user suggestions and intellectual property rights and licensing are proved to be important for differentiating positively innovative firms from non-innovative ones, as pointed by literature. Internationalization, Yesilar to the MLR analysis, proved to be an obstacle to the promotion of marketing innovation, as much as the business size. In addition, with this DA, product innovation and organizational innovation, proved to be differentiators for distinguish marketing innovative from non-innovative firms, although not significant in differentiating the level of marketing innovation in MLR.

Table 24. Linear regression model vs. discriminant analysis.

Explanatory Variables	Hypothesis	Acronyms	MLR	DA
Product Innovation	H1	INPDGD	NS	+
	H2	INPDSV	NS	+
Organizational Innovation	H3	Inov_Org	+	+
Customer and/or User Suggestions	H4	Sug_Users	+	+
Intellectual Property Rights and Licensing	H5	Prop_Intellectual	+	+
Higher Education of Employees	H6	EMPUD	NS	NS
Business Size	H7	SIZE14_COD	NS	-
Geographic Markets	H8	M_GEO	NS	NS
Internationalization	H9	SLO14	-	-
Membership of a Group of Firms	H10	GP	NS	NS

5. Conclusions

This study explored marketing innovation in Portuguese firms between 2012 and 2014. Two multivariate statistical techniques were performed to confirm the hypothesis resulting from the literature review, namely: multiple linear regression and discriminant analysis. Both had different objectives. In the first one, it was aimed to understand which factors contributed more to explain the Marketing Innovation Index or marketing innovation level of firms and in the second one it was aimed to define a profile of the firms that do not innovate and innovate in marketing.

Regarding multiple linear regression, it was concluded that the model is significant, and it explains 20.4% of the Marketing Innovation Index. Organizational Innovation Index, customer suggestions and/or users and IPRL were the variables with the greatest contribution to explain Marketing Innovation Index. In fact, about the contribution of IPRL, they can increase the capacity of Marketing Innovation in the sense that firms feel more confident in sharing knowledge because they are protected [6]. In turn, the positive contribution of organizational innovation can be explained by the fact that firms increasingly apply improvements in organizational management through innovative marketing measures [7]. Finally, the contribution of customer suggestions and/or users may be related to the fact that they are the consumers of the innovations implemented through products and/or services, so they have a good perception of what they want to buy [8].

Discriminant analysis reinforced the results obtained through multiple linear regression and proved useful to understand that the different indices of Marketing Innovation display no influence on the results, since they were equivalent when used a dummy variable (innovated/not innovated in marketing). In order to summarise the results of the discriminant analysis, the variables show little discriminative power, however, most of the 7,083 cases (both in the original classification and in the cross validation) were correctly classified. Product Innovation and Organizational Innovation, proved to be important to distinguish innovative from non-innovative in marketing firms, but not relevant to explain the increase of the level of marketing innovation.

Geographic markets, a higher academic level of the employees and belonging to a group of firms do not contribute to explain the Marketing Innovation, thus rejecting the hypothesis initially placed: H6, H8 and H10.

Internationalization, proved to be an obstacle to promotion of marketing innovation, as much as the business size, thus H7 and H9 are verified but with a sign different from expected in the literature.

This study offers some difficulties and limitations, namely that most of the existing literature on Marketing Innovation is considerably recent and that, since 2014 (date of the CIS database) to present, behavioral changes may occur in firms regarding the importance of marketing and innovation itself.

Math. Comput. Appl. **2019**, *24*, 99

The results show that there is still room for exploring the factors explaining marketing innovation, and this study took some steps in this direction. In fact, a future study may consider other variables, such as cooperation, marketing activities and/or public financial support [33], since the variables used in this study although relevant, are insufficient to fully explain Marketing Innovation. In parallel, it would be relevant to obtain more recent data through primary data, for example, firm surveys, in order to enrich and complement this study or to expand the research.

Author Contributions: Cooperative work throughout the manuscript. All authors read and approved the final manuscript.

Acknowledgments: We are grateful for ESTG—P. PORTO and CIISESI for the support in the preparation of this manuscript and in the participation in SYMCOMP 2019—4th International Conference on Numerical and Symbolic Computation. Developments and Applications.

Conflicts of Interest: The authors declare no conflict of interest.

Appendix A

Appendix A.1. Frequency Tables

Table A1. Business size | frequency table.

		Frequency	Percent	Valid Percent	Cumulative Percent
		Size14_COD			
Valid	10–49 employees	4704	66.4	74.1	74.1
	50–249 employees	1311	18.5	20.7	94.8
	>= 250 employees	332	4.7	5.2	100.0
	Total	6347	89.6	100.0	
Missing	System	736	10.4		
	Total	7083	100.0		

Table A2. Belonging to a group of firms | frequency table.

		Frequency	Percent	Valid Percent	Cumulative Percent
		GP			
Valid	No	5077	71.7	71.7	71.7
	Yes	2006	28.3	28.3	100.0
	Total	7083	100.0	100.0	

Table A3. Geographic markets | frequency table.

		Frequency	Percent	Valid Percent	Cumulative Percent
		M_GEO			
Valid	1.00	1172	16.5	16.5	16.5
	2.00	1659	23.4	23.4	40.0
	3.00	1701	24.0	24.0	64.0
	4.00	2551	36.0	36.0	100.0
	Total	7083	100.0	100.0	

Table A4. Higher education of employees | frequency table.

		Frequency	Percent	Valid Percent	Cumulative Percent
			EMPUD		
Valid	0%	1107	15.6	15.6	15.6
	1%–4%	1825	25.8	25.8	41.4
	5%–9%	929	13.1	13.1	54.5
	10%–24%	1451	20.5	20.5	75.0
	25%–49%	770	10.9	10.9	85.9
	50%–74%	495	7.0	7.0	92.9
	75%–100%	506	7.1	7.1	100.0
	Total	7083	100.0	100.0	

Table A5. Intellectual property rights and licensing | frequency table.

		Frequency	Percent	Valid Percent	Cumulative Percent
			Prop_Intellectual		
Valid	0.00	6087	85.9	85.9	85.9
	1.00	791	11.2	11.2	97.1
	2.00	155	2.2	2.2	99.3
	3.00	38	0.5	0.5	99.8
	4.00	12	0.2	0.2	100.0
	Total	7083	100.0	100.0	

Table A6. Marketing Innovation Index | frequency table.

		Frequency	Percent	Valid Percent	Cumulative Percent
			Inov_Mark		
Valid	0.00	4824	68.1	68.1	68.1
	1.00	981	13.9	13.9	82.0
	2.00	638	9.0	9.0	91.0
	3.00	358	5.1	5.1	96.0
	4.00	282	4.0	4.0	100.0
	Total	7083	100.0	100.0	

Table A7. Measures of central tendency and asymmetry and kurtosis | descriptive analysis.

						Statistics						
		GP	INPDGD	INPDSV	Inov_Mark	M_GEO	Inov_Org	Prop_Intellectual	Sug_Utilizadores	SLO14	SIZE14_COD	EMPUD
N	Valid	7083	7083	7083	7083	7083	7083	7083	4164	7083	6347	7083
	Missing	0	0	0	0	0	0	0	2919	0	736	0
Mean		0.28	0.27	0.18	0.6295	2.7950	0.5382	0.1783	6.1720	0.1617	1.31	2.35
Median		0.00	0.00	0.00	0.0000	3.0000	0.0000	0.0000	6.0000	0.0000	1.00	2.00
Mode		0	0	0	0.00	4.00	0.00	0.00	0.00	0.00	1	1
Std. Deviation		0.451	0.441	0.388	1.09296	1.10200	0.94186	0.49279	4.69351	0.29247	0.565	1.782
Skewness		0.963	1.064	1.624	1.737	-0.331	1.580	3.381	0.385	1.745	1.650	0.473
Std. Error of Skewness		0.029	0.029	0.029	0.029	0.029	0.029	0.029	0.038	0.029	0.031	0.029
Kurtosis		-1.074	-0.867	0.639	2.022	-1.256	1.152	13.876	-0.750	1.636	1.710	-0.754
Std. Error of Kurtosis		0.058	0.058	0.058	0.058	0.058	0.058	0.058	0.076	0.058	0.061	0.058
Minimum		0	0	0	0.00	1.00	0.00	0.00	0.00	0.00	1	0
Maximum		1	1	1	4.00	4.00	3.00	4.00	18.00	1.00	3	6
Percentiles	25	0.00	0.00	0.00	0.0000	2.0000	0.0000	0.0000	2.0000	0.0000	1.00	1.00
	50	0.00	0.00	0.00	0.0000	3.0000	.0000	0.0000	6.0000	0.0000	1.00	2.00
	75	1.00	1.00	0.00	1.0000	4.0000	1.0000	0.0000	10.0000	0.1700	2.00	4.00

Table A8. Organizational Innovation Index | frequency table.

		Frequency	Percent	Valid Percent	Cumulative Percent
				Inov_Org	
Valid	0.00	4996	70.5	70.5	70.5
	1.00	898	12.7	12.7	83.2
	2.00	653	9.2	9.2	92.4
	3.00	536	7.6	7.6	100.0
	Total	7083	100.0	100.0	

Table A9. Service innovation | frequency table.

		Frequência	Percent	Valid Percent	Cumulative Percent
				INPDSV	
Valid	No	5774	81.5	81.5	81.5
	Yes	1309	18.5	18.5	100.0
	Total	7083	100.0	100.0	

Table A10. Goods innovation | frequency table.

		Frequency	Percent	Valid Percent	Cumulative Percent
				INPDGD	
Valid	No	5205	73.5	73.5	73.5
	Yes	1878	26.5	26.5	100.0
	Total	7083	100.0	100.0	

Table A11. Customer and/or user suggestions | frequency table.

		Frequency	Percent	Valid Percent	Cumulative Percent
				Sug_Users	
Valid	0.00	730	10.3	17.5	17.5
	1.00	101	1.4	2.4	20.0
	2.00	276	3.9	6.6	26.6
	3.00	356	5.0	8.5	35.1
	4.00	277	3.9	6.7	41.8
	5.00	218	3.1	5.2	47.0
	6.00	352	5.0	8.5	55.5
	7.00	256	3.6	6.1	61.6
	8.00	266	3.8	6.4	68.0
	9.00	256	3.6	6.1	74.2
	10.00	211	3.0	5.1	79.2
	11.00	188	2.7	4.5	83.7
	12.00	251	3.5	6.0	89.8
	13.00	142	2.0	3.4	93.2
	14.00	97	1.4	2.3	95.5
	15.00	71	1.0	1.7	97.2
	16.00	34	0.5	0.8	98.0
	17.00	34	0.5	0.8	98.8
	18.00	48	0.7	1.2	100.0
	Total	4164	58.8	100.0	
Omisso	System	2919	41.2		
Total		7083	100.0		

Appendix A.2. Graphics

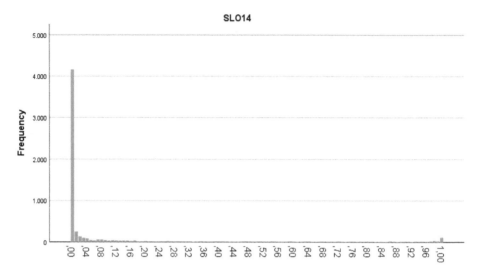

Figure A1. Internationalization | bar chart.

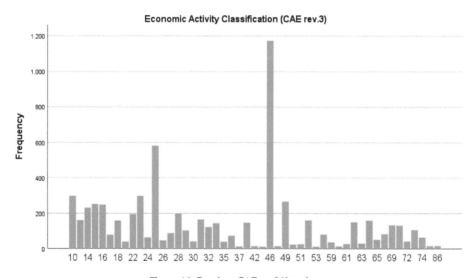

Figure A2. Bar chart CAE rev.3 | bar chart.

Appendix A.3. Multiple Linear Regression | Tables

Table A12. Coefficients | linear regression by forward method.

		Coefficients [a]					
	Model	Unstandardized Coefficients		Standardized Coefficients	t	Sig.	
		B	Std. Error	Beta			
1	(Constant)	0.308	0.072		4.271	0.000	
	GP	−0.031	0.045	−0.011	−0.675	0.500	
	M_GEO	0.018	0.019	0.015	0.937	0.349	
	Inov_Org	0.302	0.018	0.259	16.512	0.000	
	Prop_Intellectual	0.324	0.034	0.145	9.438	0.000	
	Sug_Utilizadores	0.046	0.004	0.172	10.827	0.000	
	SLO14	−0.294	0.070	−0.069	−4.183	0.000	
	SIZE14_COD	−0.029	0.033	−0.014	−0.865	0.387	
	EMPUD	0.001	0.012	0.002	0.128	0.898	
	INPDGD	0.223	0.040	0.089	5.562	0.000	
	INPDSV	0.240	0.043	0.089	5.572	0.000	

[a] Dependent Variable: Inov_Mark.

Table A13. Variables entered/removed | Stepwise.

		Variables Entered/Removed [a]	
Model	Variables Entered	Variables Removed	Method
1	Inov_Org	-	Stepwise (Criteria: Probability-of-F-to-enter <= 0.050. Probability-of-F-to-remove >= 0.100).
2	Sug_Users	-	Stepwise (Criteria: Probability-of-F-to-enter <= 0.050. Probability-of-F-to-remove >= 0.100).
3	Prop_Intellectual	-	Stepwise (Criteria: Probability-of-F-to-enter <= 0.050. Probability-of-F-to-remove >= 0.100).
4	INPDSV	-	Stepwise (Criteria: Probability-of-F-to-enter <= 0.050. Probability-of-F-to-remove >= 0.100).
5	INPDGD	-	Stepwise (Criteria: Probability-of-F-to-enter <= 0.050. Probability-of-F-to-remove >= 0.100).
6	SLO14	-	Stepwise (Criteria: Probability-of-F-to-enter <= 0.050. Probability-of-F-to-remove >= 0.100).

[a] Dependent Variable: Inov_Mark.

References

1. Oslo Manual—Guidelines for Collecting and Interpreting Innovation Data. Available online: https://www.oecd-ilibrary.org/science-and-technology/oslo-manual_9789264013100-en (accessed on 20 November 2019).
2. Geldes, C.; Felzensztein, C.; Palacios-Fenech, J. Technological and non-technological innovations, performance and propensity to innovate across industries: The case of an emerging economy. *Ind. Mark. Manag.* **2017**, *61*, 55–66. [CrossRef]
3. Mendonça, S.; Pereira, T.S.; Godinho, M.M. Trademarks as an indicator of innovation and industrial change. *Res. Policy* **2004**, *33*, 1385–1404. [CrossRef]
4. Matte, J.; Graebin, R.E. Inovação de Marketing na Perspectiva Literária. *Tópicos de Marketing* **2017**, *2*, 258.
5. Correia, A.; Braga, A.; Braga, V. Innovation in Portuguese Firms, Using CIS2010. *WSEAS Trans. Bus. Econ.* **2017**, *14*, 55–63.
6. Olaisen, J.; Revang, O. The dynamics of intellectual property rights for trust, knowledge sharing and innovation in project teams. *Int. J. Inf. Manag.* **2017**, *37*, 583–589. [CrossRef]
7. Fleacă, E.; Fleacă, B.; Maiduc, S. Fostering Organizational Innovation based on modeling the marketing research process through event-driven process chain (EPC). *TEM J.* **2016**, *5*, 460–466.

8. Rebane, T. *Complementarities in Performance Between Product Innovation, Marketing Innovation and Cooperation with Clients*; The University of Tartu: Tartu, Estonia, 2018.

9. Kotler, P.; Kartajaya, H.; Setiawan, I. *Marketing 4.0: Moving from Traditional to Digital*; Review; John Wiley & Sons: Hoboken, NJ, USA, 2016; pp. 1–3.

10. American Marketing Association. New Definition for Marketing. Press Release, 1–3. Available online: https://www.ama.org/the-definition-of-marketing-what-is-marketing/ (accessed on 20 November 2019).

11. Kotler, P.; Armstrong, G. Principles of Marketing Channel Management. *J. Mark.* **1978**, *42*, 105.

12. Dantas, J.; Moreira, A.C. *O Processo de Inovação*; Lidel: Lisbon, Portugal, 2011.

13. Dalfovo, M.S.; Hoffmann, V.E.; Lazzarotti, F. A Bibliometric Study of Innovation Based on Schumpeter. *J. Tech. Manag. Innov.* **2011**, *6*, 121–135.

14. Drucker, P. *Innovation and Entrepeurnship*; Harper & Row, Publishers, Inc.: New York, NY, USA, 1985.

15. Leonard, D.; Walter, S. *When Sparks Fly*; Harvard Business School Press: Boston, MA, USA, 1999; Volume 7.

16. Barack Obama quoted in Business Week's "In" subsection, p. 6, November 2007. Available online: https://www.freshconsulting.com/what-is-innovation/ (accessed on 20 November 2019).

17. Razeghi, A. *The Riddle: Where Ideas Come from and How to Have Better Ones*; John Wiley & Sons: Hoboken, NJ, USA, 2008.

18. Martin, G. The importance of marketing innovation in new economy. *Am. J. Bus. Educ.* **2011**, *4*, 15.

19. Grimpe, C.; Sofka, W.; Bhargava, M.; Chatterjee, R. R&D, Marketing Innovation, and New Product Performance: A Mixed Methods Study. *J. Prod. Innov. Manag.* **2017**, *34*, 360–383.

20. INNOREGIO: Dissemination of innovation and knowledge management techniques. Available online: https://www.urenio.org/tools/en/marketing_of_innovation.pdf (accessed on 20 November 2019).

21. Gupta, A.K.; Raj, S.P.; Wilemon, D. A Model for Studying R&D. Marketing Interface in the Product Innovation Process. *J. Mark.* **1986**, *50*, 7–17.

22. Junge, M.; Severgnini, B.; Sørensen, A. Product-Marketing Innovation, Skills, and Firm Productivity. No. 01-2012. Available online: https://ideas.repec.org/p/hhs/cbsnow/2012_001.html (accessed on 22 November 2019).

23. Ganzer, P.P.; Chais, C.; Olea, P.M. Product, process, marketing and organizational innovation in industries of the flat knitting sector. *RAI* **2017**, *14*, 321–332. [CrossRef]

24. Higgins, J.M. *Innovate or Evaporate: Test and Improve Your Organization's Innovation Quotient*; New Management: Winter Park, FL, USA, 1995.

25. Kilinc, N.; Ozturk, G.B.; Yitmen, I. The changing role of the client in driving innovation for design-build projects: stakeholders' perspective. *Procedia Econ. Financ.* **2015**, *21*, 279–287. [CrossRef]

26. Cabigiosu, A.; Campagnolo, D. Innovation and growth in KIBS: The role of clients' collaboration and service customisation. *Ind. Innov.* **2019**, *26*, 592–618. [CrossRef]

27. Tauber, E.M. How market research discourages major innovation. *Bus. Horiz.* **1974**, *17*, 22–26. [CrossRef]

28. Correia, A.; Machado, A.; Braga, A. Marketing Innovation Using CIS Portuguese Dataset. *Int. J. Ecol. Stat.* **2017**, *5*, 27–33.

29. Junge, M.; Severgnini, B.; Sørensen, A. Product-Marketing Innovation, Skills, and Firm Productivity Growth. *Rev. Income Wealth* **2016**, *62*, 724–757. [CrossRef]

30. Sok, P.; O'Cass, A.; Sok, K.M. Achieving superior SME performance: Overarching role of marketing, innovation, and learning capabilities. *Australas. Mark. J.* **2013**, *21*, 161–167. [CrossRef]

31. Ajayi, O.M.; Morton, S.C. Exploring the enablers of organizational and marketing innovations in SMEs: Findings from south-western Nigeria. *SAGE Open* **2015**, *5*, 1–13. [CrossRef]

32. Ungerman, O.; Dedkova, J.; Gurinova, K. The impact of marketing innovation on the competitiveness of enterprises in the context of industry 4.0. *J. Competi.* **2018**, *10*, 132. [CrossRef]

33. Moreira, J.R.M. *Inovação de Marketing: Estudo dos Factores Determinantes da Capacidade Inovadora de Marketing das Empresas Portuguesas*. Ph.D. Thesis, Universidade da Beira Interior, Covilhã, Portugal, 2010.

34. Haucap, J. Merger Effects on Innovation: A Rationale for Stricter Merger Control? Available online: https://www.econstor.eu/handle/10419/168579 (accessed on 20 November 2019).

35. Entezarkheir, M.; Moshiri, S. Mergers and innovation: Evidence from a panel of US firms. *Econ. Innov. New Tech.* **2018**, *27*, 132–153. [CrossRef]

36. Wiechoczek, J. Marketing innovations and modernisation of value for customer in the market of high technology products. *Handel Wewnętrzny* **2016**, *363*, 338–349.

37. Crick, D.; Crick, J. The first export order: A marketing innovation revisited. *J. Strateg. Mark.* **2015**, *24*, 77–89. [CrossRef]
38. Braga, A.; Braga, V. Factors influencing innovation decision making in Portuguese firms. *IJIL* **2013**, *14*, 329–349. [CrossRef]
39. Coelho, P.P. Decreto-Lei n.° 98/2015 de 2 de junho do Ministério das Finanças. Diário Da República: I Série, 3470–3493. Available online: https://dre.pt/application/file/67361214 (accessed on 20 November 2019).
40. Constantin, C. Using the Regression Model in multivariate data analysis. *Bull. Transilvania Univ. Brasov. Econ. Sci.* **2017**, *10*, 27–34.
41. Hall, A.; Neves, C.; Pereira, A. *Grande Maratona de Estatística no SPSS*; Escolar Editora: Lisboa, Portugal, 2011.
42. Woodward, E.C., Jr.; Drew, W.B. Flame-quenching device using nonparallel walls. *Combust. Flame* **1971**, *16*, 203–204. [CrossRef]
43. Sarstedt, M.; Mooi, E. *A Concise Guide to Market Research*; Springer-Verlag: Berlin/Heidelberg, Germany, 2014.
44. Maroco, J. *Análise Estatística—Com a Utilização do SPSS*, 3rd ed.; Edições Sílabo: Lisboa, Portugal, 2010.
45. Pereira, A. *SPSS: Guia Prático de Utilização: Análise de Dados Para Ciências Sociais e Psicologia*; Edições Silabo: Lisboa, Portugal, 2003.
46. Derriennic, Y.; Lin, M. The central limit theorem. *Probab. Theory Relat. Fields* **2003**, *125*, 73–76. [CrossRef]
47. Favero, N. Revisiting Multicollinearity: When Correlated Predictors Exhibit Nonlinear Effects or Contain Measurement Error. Available online: https://appam.confex.com/appam/2016/webprogram/Paper17653.html (accessed on 20 November 2019).

Mathematical and Computational Applications

Article

Numerical Optimal Control of HIV Transmission in Octave/MATLAB

Carlos Campos [1,†], Cristiana J. Silva [2,*,†] and Delfim F. M. Torres [2,†]

1   Department of Mathematics, Polytechnic of Leiria, 2411-901 Leiria, Portugal; carlos.campos@ipleiria.pt
2   Center for Research and Development in Mathematics and Applications (CIDMA),
     Department of Mathematics, University of Aveiro, 3810-193 Aveiro, Portugal; delfim@ua.pt
*   Correspondence: cjoaosilva@ua.pt
†   These authors contributed equally to this work.

Received: 1 October 2019; Accepted: 17 December 2019; Published: 19 December 2019

Abstract: We provide easy and readable GNU Octave/MATLAB code for the simulation of mathematical models described by ordinary differential equations and for the solution of optimal control problems through Pontryagin's maximum principle. For that, we consider a normalized HIV/AIDS transmission dynamics model based on the one proposed in our recent contribution (Silva, C.J.; Torres, D.F.M. A SICA compartmental model in epidemiology with application to HIV/AIDS in Cape Verde. *Ecol. Complex.* **2017**, *30*, 70–75), given by a system of four ordinary differential equations. An HIV initial value problem is solved numerically using the ode45 GNU Octave function and three standard methods implemented by us in Octave/MATLAB: Euler method and second-order and fourth-order Runge–Kutta methods. Afterwards, a control function is introduced into the normalized HIV model and an optimal control problem is formulated, where the goal is to find the optimal HIV prevention strategy that maximizes the fraction of uninfected HIV individuals with the least HIV new infections and cost associated with the control measures. The optimal control problem is characterized analytically using the Pontryagin Maximum Principle, and the extremals are computed numerically by implementing a forward-backward fourth-order Runge–Kutta method. Complete algorithms, for both uncontrolled initial value and optimal control problems, developed under the free GNU Octave software and compatible with MATLAB are provided along the article.

Keywords: numerical algorithms; optimal control; HIV/AIDS model; GNU Octave; open source code for optimal control through Pontryagin Maximum Principle

MSC: 34K28; 49N90; 92D30

1. Introduction

In recent years, mathematical modeling of processes in biology and medicine, in particular in epidemiology, has led to significant scientific advances both in mathematics and biosciences [1,2]. Applications of mathematics in biology are completely opening new pathways of interactions, and this is certainly true in the area of optimal control: a branch of applied mathematics that deals with finding control laws for dynamical systems over a period of time such that an objective functional is optimized [3,4]. It has numerous applications in both biology and medicine [5–8].

To find the best possible control for taking a dynamical system from one state to another, one uses, in optimal control theory, the celebrated Pontryagin's maximum principle (PMP), formulated in 1956 by the Russian mathematician Lev Pontryagin and his collaborators [3]. Roughly speaking, PMP states that it is necessary for any optimal control along with the optimal state trajectory to satisfy the so-called Hamiltonian system, which is a two-point boundary value problem, plus a maximality condition on the Hamiltonian. Although a classical result, PMP is usually not taught to biologists and mathematicians working on mathematical biology. Here, we show how such scientists can easily implement the necessary optimality conditions given by the PMP, numerically, and can benefit from the power of optimal control theory. For that, we consider a mathematical model for HIV.

HIV modeling and optimal control is a subject under strong current research: see, e.g., Reference [9] and the references therein. Here, we consider the SICA epidemic model for HIV transmission proposed in References [10,11], formulate an optimal control problem with the goal to find the optimal HIV prevention strategy that maximizes the fraction of uninfected HIV individuals with least HIV new infections and cost associated with the control measures, and give complete algorithms in GNU Octave to solve the considered problems. We trust that our work, by providing the algorithms in an open programming language, contributes to reducing the so-called "replication crisis" (an ongoing methodological crisis in which it has been found that many scientific studies are difficult or impossible to replicate or reproduce [12]) in the area of optimal biomedical research. We trust our current work will be very useful to a practitioner from the disease control area and will become a reference in the field of epidemiology for those interested to include an optimal control component in their work.

2. A Normalized SICA HIV/AIDS Model

We consider the SICA epidemic model for HIV transmission proposed in References [10,11], which is given by the following system of ordinary differential equations:

$$\begin{cases} S'(t) = bN(t) - \lambda(t)S(t) - \mu S(t) \\ I'(t) = \lambda(t)S(t) - (\rho + \phi + \mu)I(t) + \alpha A(t) + \omega C(t) \\ C'(t) = \phi I(t) - (\omega + \mu)C(t) \\ A'(t) = \rho I(t) - (\alpha + \mu + d)A(t). \end{cases} \tag{1}$$

The model in Equation (1) subdivides human population into four mutually exclusive compartments: susceptible individuals (S); HIV-infected individuals with no clinical symptoms of AIDS (the virus is living or developing in the individuals but without producing symptoms or only mild ones) but able to transmit HIV to other individuals (I); HIV-infected individuals under ART treatment (the so-called chronic stage) with a viral load remaining low (C); and HIV-infected individuals with AIDS clinical symptoms (A). The total population at time t, denoted by $N(t)$, is given by $N(t) = S(t) + I(t) + C(t) + A(t)$. Effective contact with people infected with HIV is at a rate $\lambda(t)$, given by

$$\lambda(t) = \frac{\beta}{N(t)} \left(I(t) + \eta_C C(t) + \eta_A A(t) \right),$$

where β is the effective contact rate for HIV transmission. The modification parameter $\eta_A \geq 1$ accounts for the relative infectiousness of individuals with AIDS symptoms in comparison to those infected with HIV with no AIDS symptoms (individuals with AIDS symptoms are more infectious than HIV-infected individuals—pre-AIDS). On the other hand, $\eta_C \leq 1$ translates the partial restoration of immune function of individuals with HIV infection that use ART correctly [10]. All individuals suffer from natural death at a constant rate μ. Both HIV-infected individuals with and without AIDS symptoms have access to ART treatment: HIV-infected individuals with no AIDS symptoms I progress to the class of individuals with HIV infection under ART treatment C at a rate ϕ, and HIV-infected individuals with AIDS symptoms are treated for HIV at rate α. An HIV-infected individual with AIDS symptoms A that starts treatment moves to the class of HIV-infected individuals I and after, if the treatment is maintained, will be transferred to the chronic class C. Individuals in the class C leave to the class I at a rate ω due to a default treatment. HIV-infected individuals with no AIDS symptoms I that do not take ART treatment progress to the AIDS class A at rate ρ. Finally, HIV-infected individuals with AIDS symptoms A suffer from an AIDS-induced death at a rate d.

In the situation where the total population size $N(t)$ is not constant, it is often convenient to consider the proportions of each compartment of individuals in the population, namely

$$s = S/N, \quad i = I/N, \quad c = C/N, \quad r = R/N.$$

The state variables s, i, c, and a satisfy the following system of differential equations:

$$\begin{cases} s'(t) = b(1 - s(t)) - \beta(i(t) + \eta_C c(t) + \eta_A a(t))s(t) + d\,a(t)\,s(t) \\ i'(t) = \beta\,(i(t) + \eta_C\,c(t) + \eta_A a(t))\,s(t) - (\rho + \phi + b)i(t) + \alpha a(t) + \omega c(t) + d\,a(t)\,i(t) \\ c'(t) = \phi i(t) - (\omega + b)c(t) + d\,a(t)\,c(t) \\ a'(t) = \rho\,i(t) - (\alpha + b + d)a(t) + d\,a^2(t) \end{cases} \qquad (2)$$

with $s(t) + i(t) + c(t) + a(t) = 1$ for all $t \in [0, T]$.

3. Numerical Solution of the SICA HIV/AIDS Model

In this section, we consider Equation (2) subject to the initial conditions given by

$$s(0) = 0.6, \quad i(0) = 0.2, \quad c(0) = 0.1, \quad a(0) = 0.1, \qquad (3)$$

by the fixed parameter values from Table 1, and by the final time value of $T = 20$ (years).

Table 1. Parameter values of the HIV/AIDS model in Equation (2) taken from Reference [11] and references cited therein.

Symbol	Description	Value
μ	Natural death rate	1/69.54
b	Recruitment rate	2.1μ
β	HIV transmission rate	1.6
η_C	Modification parameter	0.015
η_A	Modification parameter	1.3
ϕ	HIV treatment rate for I individuals	1
ρ	Default treatment rate for I individuals	0.1
α	AIDS treatment rate	0.33
ω	Default treatment rate for C individuals	0.09
d	AIDS induced death rate	1

All our algorithms, developed to solve numerically the initial value problems in Equations (2) and (3), are developed under the free GNU Octave software (version 5.1.0), a high-level programming language primarily intended for numerical computations and is the major free and mostly compatible alternative to MATLAB [13]. We implement three standard basic numerical techniques: Euler, second-order Runge–Kutta, and fourth-order Runge–Kutta. We compare the obtained solutions with the one obtained using the ode45 GNU Octave function.

3.1. Default ode45 Routine of GNU Octave

Using the provided ode45 function of GNU Octave, which solves a set of non-stiff ordinary differential equations with the well known explicit Dormand–Prince method of order 4, one can solve the initial value problems in Equations (2) and (3) as follows:

```
function dy = odeHIVsystem(t,y)
% Parameters of the model
mi = 1.0 / 69.54; b = 2.1 * mi; beta = 1.6;
etaC = 0.015; etaA = 1.3; fi = 1.0; ro = 0.1;
```

```
alfa = 0.33; omega = 0.09; d = 1.0;

% Differential equations of the model
dy = zeros(4,1);
aux1 = beta * (y(2) + etaC * y(3) + etaA * y(4)) * y(1);
aux2 = d * y(4);
dy(1) = b * (1 - y(1)) - aux1 + aux2 * y(1);
dy(2) = aux1 - (ro + fi + b - aux2) * y(2) + alfa * y(4) + omega * y(3);
dy(3) = fi * y(2) - (omega + b - aux2) * y(3);
dy(4) = ro * y(2) - (alfa + b + d - aux2) * y(4);
```

On the GNU Octave interface, one should then type the following instructions:

```
>> T = 20; N = 100;
>> [vT,vY] = ode45(@odeHIVsystem,[0:T/N:T],[0.6 0.2 0.1 0.1]);
```

Next, we show how such approach compares with standard numerical techniques.

3.2. Euler's Method

Given a well-posed initial-value problem

$$\frac{dy}{dt} = f(t,y) \quad \text{with} \quad y(a) = \alpha \quad \text{and} \quad a \le t \le b,$$

Euler's method constructs a sequence of approximation points $(t,w) \approx (t, y(t))$ to the exact solution of the ordinary differential equation by $t_{i+1} = t_i + h$ and $w_{i+1} = w_i + hf(t_i, w_i)$, $i = 0, 1, \ldots, N-1$, where $t_0 = a$, $w_0 = \alpha$, and $h = (b-a)/N$. Let us apply Euler's method to approximate each one of the four state variables of the system of ordinary differential equations (Equation (2)). Our odeEuler GNU Octave implementation is as follows:

```
function dy = odeEuler(T)
% Parameters of the model
mi = 1.0 / 69.54; b = 2.1 * mi; beta = 1.6;
etaC = 0.015; etaA = 1.3; fi = 1.0; ro = 0.1;
alfa = 0.33; omega = 0.09; d = 1.0;

% Parameters of the Euler method
test = -1; deltaError = 0.001; M = 100;
t = linspace(0,T,M+1); h = T / M;
S = zeros(1,M+1); I = zeros(1,M+1);
C = zeros(1,M+1); A = zeros(1,M+1);

% Initial conditions of the model
S(1) = 0.6; I(1) = 0.2; C(1) = 0.1; A(1) = 0.1;

% Iterations of the method
while(test < 0)
  oldS = S; oldI = I; oldC = C; oldA = A;
  for i = 1:M
    % Differential equations of the model
    aux1 = beta * (I(i) + etaC * C(i) + etaA * A(i)) * S(i);
    aux2 = d * A(i);

    auxS = b * (1 - S(i)) - aux1 + aux2 * S(i);
    auxI = aux1 - (ro + fi + b - aux2) * I(i) + alfa * A(i) + omega * C(i);
    auxC = fi * I(i) - (omega + b - aux2) * C(i);
```

```
    auxA = ro * I(i) - (alfa + b + d - aux2) * A(i);

    % Euler new approximation
    S(i+1) = S(i) + h * auxS;
    I(i+1) = I(i) + h * auxI;
    C(i+1) = C(i) + h * auxC;
    A(i+1) = A(i) + h * auxA;
  end

  % Absolute error for convergence
  temp1 = deltaError * sum(abs(S)) - sum(abs(oldS - S));
  temp2 = deltaError * sum(abs(I)) - sum(abs(oldI - I));
  temp3 = deltaError * sum(abs(C)) - sum(abs(oldC - C));
  temp4 = deltaError * sum(abs(A)) - sum(abs(oldA - A));
  test = min(temp1,min(temp2,min(temp3,temp4)));
end
dy(1,:)  = t; dy(2,:)  = S; dy(3,:)  = I;
dy(4,:)  = C; dy(5,:)  = A;
```

Figure 1 shows the solution of the system of ordinary differential equations (Equation (2)) with the initial conditions (Equation (3)), computed by the ode45 GNU Octave function (dashed line) *versus* the implemented Euler's method (solid line). As depicted, Euler's method, although being the simplest method, gives a very good approximation to the behaviour of each of the four system variables. Both implementations use the same discretization knots in the interval $[0, T]$ with a step size given by $h = T/100$.

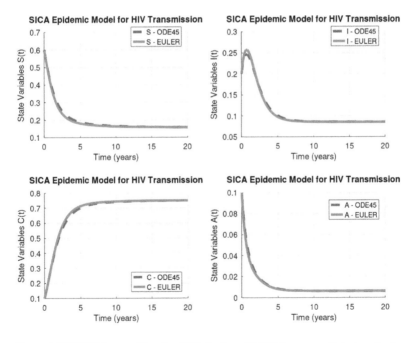

Figure 1. HIV/AIDS system (Equation (2)) behaviour: GNU Octave versus Euler's method.

Euler's method has a global error (total accumulated error) of $O(h)$, and therefore, the error bound depends linearly on the step size h, which implies that the error is expected to grow in no worse than a linear manner. Consequently, diminishing the step size should give correspondingly greater accuracy to the approximations. Table 2 lists the norm of the difference vector, where each component of this vector is the absolute difference between the results obtained by the ode45 GNU Octave function and our implementation of Euler's method, calculated by the vector norms 1, 2, and ∞.

Table 2. Norms 1, 2, and ∞ of the difference vector between ode45 GNU Octave and Euler's results.

System Variables	$S(t)$	$I(t)$	$C(t)$	$A(t)$
$\|Octave - Euler\|_1$	0.4495660	0.1646710	0.5255950	0.0443340
$\|Octave - Euler\|_2$	0.0659270	0.0301720	0.0783920	0.0101360
$\|Octave - Euler\|_\infty$	0.0161175	0.0113068	0.0190621	0.0041673

3.3. Runge–Kutta of Order Two

Given a well-posed initial-value problem, the Runge–Kutta method of order two constructs a sequence of approximation points $(t, w) \approx (t, y(t))$ to the exact solution of the ordinary differential equation by $t_{i+1} = t_i + h$, $K_1 = f(t_i, w_i)$, $K_2 = f(t_{i+1}, w_i + hK_1)$, and $w_{i+1} = w_i + h\dfrac{K_1 + K_2}{2}$, for each $i = 0, 1, \ldots, N-1$, where $t_0 = a$, $w_0 = \alpha$, and $h = (b-a)/N$. Our GNU Octave implementation of the Runge–Kutta method of order two applies the above formulation to approximate each of the four variables of the system in Equation (2). We implement the odeRungeKutta_order2 function through the following GNU Octave instructions:

```
function dy = odeRungeKutta_order2(T)
% Parameters of the model
mi = 1.0 / 69.54; b = 2.1 * mi; beta = 1.6;
etaC = 0.015; etaA = 1.3; fi = 1.0; ro = 0.1;
alfa = 0.33; omega = 0.09; d = 1.0;

% Parameters of the Runge-Kutta (2nd order) method
test = -1; deltaError = 0.001; M = 100;
t = linspace(0,T,M+1); h = T / M; h2 = h / 2;
S = zeros(1,M+1); I = zeros(1,M+1);
C = zeros(1,M+1); A = zeros(1,M+1);

% Initial conditions of the model
S(1) = 0.6; I(1) = 0.2; C(1) = 0.1; A(1) = 0.1;

% Iterations of the method
while(test < 0)
  oldS = S; oldI = I; oldC = C; oldA = A;
  for i = 1:M
    % Differential equations of the model
    % First Runge-Kutta parameter
    aux1 = beta * (I(i) + etaC * C(i) + etaA * A(i)) * S(i);
    aux2 = d * A(i);
    auxS1 = b * (1 - S(i)) - aux1 + aux2 * S(i);
    auxI1 = aux1 - (ro + fi + b - aux2) * I(i) + alfa * A(i) + omega * C(i);
    auxC1 = fi * I(i) - (omega + b - aux2) * C(i);
    auxA1 = ro * I(i) - (alfa + b + d - aux2) * A(i);
```

```
% Second Runge-Kutta parameter
auxS = S(i) + h * auxS1; auxI = I(i) + h * auxI1;
auxC = C(i) + h * auxC1; auxA = A(i) + h * auxA1;
aux1 = beta * (auxI + etaC * auxC + etaA * auxA) * auxS;
aux2 = d * auxA;
auxS2 = b * (1 - auxS) - aux1 + aux2 * auxS;
auxI2 = aux1 - (ro + fi + b - aux2) * auxI + alfa * auxA + omega * auxC;
auxC2 = fi * auxI - (omega + b - aux2) * auxC;
auxA2 = ro * auxI - (alfa + b + d - aux2) * auxA;
% Runge-Kutta new approximation
S(i+1) = S(i) + h2 * (auxS1 + auxS2);
I(i+1) = I(i) + h2 * (auxI1 + auxI2);
C(i+1) = C(i) + h2 * (auxC1 + auxC2);
A(i+1) = A(i) + h2 * (auxA1 + auxA2);
end
% Absolute error for convergence
temp1 = deltaError * sum(abs(S)) - sum(abs(oldS - S));
temp2 = deltaError * sum(abs(I)) - sum(abs(oldI - I));
temp3 = deltaError * sum(abs(C)) - sum(abs(oldC - C));
temp4 = deltaError * sum(abs(A)) - sum(abs(oldA - A));
test = min(temp1,min(temp2,min(temp3,temp4)));
end
dy(1,:)  = t; dy(2,:)  = S; dy(3,:)  = I;  dy(4,:)  = C; dy(5,:)  = A;
```

Figure 2 shows the solution of the system of Equation (2) with the initial value conditions in Equation (3) computed by the ode45 GNU Octave function (dashed line) *versus* our implementation of the Runge–Kutta method of order two (solid line). As we can see, Runge–Kutta's method produces a better approximation than Euler's method, since both curves in each plot of Figure 2 are indistinguishable.

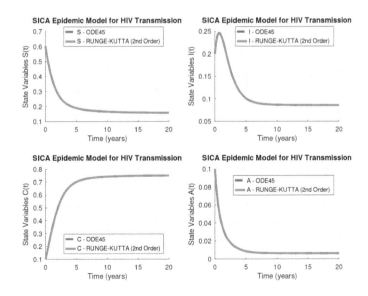

Figure 2. HIV/AIDS system (Equation (2)): GNU Octave *versus* Runge–Kutta's method of order two.

Runge–Kutta's method of order two (RK2) has a global truncation error of order $O\left(h^2\right)$, and as it is known, this truncation error at a specified step measures the amount by which the exact solution to the differential equation fails to satisfy the difference equation being used for the approximation at that step. This might seems like an unlikely way to compare the error of various methods, since we really want to know how well the approximations generated by the methods satisfy the differential equation, not the other way around. However, we do not know the exact solution, so we cannot generally determine this, and the truncation error will serve quite well to determine not only the error of a method but also the actual approximation error. Table 3 lists the norm of the difference vector between the results from ode45 routine and Runge–Kutta's method of order two results.

Table 3. Norms 1, 2, and ∞ of the difference vector between ode45 GNU Octave and RK2 results.

System Variables	$S(t)$	$I(t)$	$C(t)$	$A(t)$
$\|Octave - RungeKutta2\|_1$	0.0106530	0.0105505	0.0151705	0.0044304
$\|Octave - RungeKutta2\|_2$	0.0014868	0.0025288	0.0022508	0.0011695
$\|Octave - RungeKutta2\|_\infty$	0.0003341	0.0009613	0.0006695	0.0004678

3.4. Runge–Kutta of Order Four

The Runge–Kutta method of order four (RK4) constructs a sequence of approximation points $(t, w) \approx (t, y(t))$ to the exact solution of an ordinary differential equation by $t_{i+1} = t_i + h$, $K_1 = f(t_i, w_i)$, $K_2 = f\left(t_i + \frac{h}{2}, w_i + \frac{h}{2}K_1\right)$, $K_3 = f\left(t_i + \frac{h}{2}, w_i + \frac{h}{2}K_2\right)$, $K_4 = f(t_{i+1}, w_i + hK_3)$, and $w_{i+1} = w_i + \frac{h}{6}(K_1 + 2K_2 + 2K_3 + K_4)$, for each $i = 0, 1, \ldots, N - 1$, where $t_0 = a$, $w_0 = \alpha$, and $h = (b - a)/N$. Our GNU Octave implementation of the Runge–Kutta method of order four applies the above formulation to approximate the solution of the system in Equation (2) with the initial conditions of Equation (3) through the following instructions:

```
function dy = odeRungeKutta_order4(T)
% Parameters of the model
mi = 1.0 / 69.54; b = 2.1 * mi; beta = 1.6;
etaC = 0.015; etaA = 1.3; fi = 1.0; ro = 0.1;
alfa = 0.33; omega = 0.09; d = 1.0;

% Parameters of the Runge-Kutta (4th order) method
test = -1; deltaError = 0.001; M = 100;
t = linspace(0,T,M+1);
h = T / M; h2 = h / 2; h6 = h / 6;
S = zeros(1,M+1); I = zeros(1,M+1);
C = zeros(1,M+1); A = zeros(1,M+1);

% Initial conditions of the model
S(1) = 0.6; I(1) = 0.2; C(1) = 0.1; A(1) = 0.1;
% Iterations of the method
while(test < 0)
    oldS = S; oldI = I; oldC = C; oldA = A;
    for i = 1:M
        % Differential equations of the model
        % First Runge-Kutta parameter
        aux1 = beta * (I(i) + etaC * C(i) + etaA * A(i)) * S(i);
        aux2 = d * A(i);
        auxS1 = b * (1 - S(i)) - aux1 + aux2 * S(i);
        auxI1 = aux1 - (ro + fi + b - aux2) * I(i) + alfa * A(i) + omega * C(i);
```

```
        auxC1 = fi * I(i) - (omega + b - aux2) * C(i);
        auxA1 = ro * I(i) - (alfa + b + d - aux2) * A(i);

        % Second Runge-Kutta parameter
        auxS = S(i) + h2 * auxS1; auxI = I(i) + h2 * auxI1;
        auxC = C(i) + h2 * auxC1; auxA = A(i) + h2 * auxA1;
        aux1 = beta * (auxI + etaC * auxC + etaA * auxA) * auxS;
        aux2 = d * auxA;

        auxS2 = b * (1 - auxS) - aux1 + aux2 * auxS;
        auxI2 = aux1 - (ro + fi + b - aux2) * auxI + alfa * auxA + omega * auxC;
        auxC2 = fi * auxI - (omega + b - aux2) * auxC;
        auxA2 = ro * auxI - (alfa + b + d - aux2) * auxA;

        % Fird Runge-Kutta parameter
        auxS = S(i) + h2 * auxS2; auxI = I(i) + h2 * auxI2;
        auxC = C(i) + h2 * auxC2; auxA = A(i) + h2 * auxA2;
        aux1 = beta * (auxI + etaC * auxC + etaA * auxA) * auxS;
        aux2 = d * auxA;

        auxS3 = b * (1 - auxS) - aux1 + aux2 * auxS;
        auxI3 = aux1 - (ro + fi + b - aux2) * auxI + alfa * auxA + omega * auxC;
        auxC3 = fi * auxI - (omega + b - aux2) * auxC;
        auxA3 = ro * auxI - (alfa + b + d - aux2) * auxA;

        % Fourth Runge-Kutta parameter
        auxS = S(i) + h * auxS3; auxI = I(i) + h * auxI3;
        auxC = C(i) + h * auxC3; auxA = A(i) + h * auxA3;
        aux1 = beta * (auxI + etaC * auxC + etaA * auxA) * auxS;
        aux2 = d * auxA;

        auxS4 = b * (1 - auxS) - aux1 + aux2 * auxS;
        auxI4 = aux1 - (ro + fi + b - aux2) * auxI + alfa * auxA + omega * auxC;
        auxC4 = fi * auxI - (omega + b - aux2) * auxC;
        auxA4 = ro * auxI - (alfa + b + d - aux2) * auxA;

        % Runge-Kutta new approximation
        S(i+1) = S(i) + h6 * (auxS1 + 2 * (auxS2 + auxS3) + auxS4);
        I(i+1) = I(i) + h6 * (auxI1 + 2 * (auxI2 + auxI3) + auxI4);
        C(i+1) = C(i) + h6 * (auxC1 + 2 * (auxC2 + auxC3) + auxC4);
        A(i+1) = A(i) + h6 * (auxA1 + 2 * (auxA2 + auxA3) + auxA4);
    end

    % Absolute error for convergence
    temp1 = deltaError * sum(abs(S)) - sum(abs(oldS - S));
    temp2 = deltaError * sum(abs(I)) - sum(abs(oldI - I));
    temp3 = deltaError * sum(abs(C)) - sum(abs(oldC - C));
    temp4 = deltaError * sum(abs(A)) - sum(abs(oldA - A));
    test = min(temp1,min(temp2,min(temp3,temp4)));
  end
  dy(1,:)  = t; dy(2,:)  = S; dy(3,:)  = I;
  dy(4,:)  = C; dy(5,:)  = A;
```

Figure 3 shows the solution of the initial value problem in Equations (2) and (3) computed by the ode45 GNU Octave function (dashed line) *versus* our implementation of the Runge–Kutta method of order four (solid line). The results of the Runge–Kutta method of order four are extremely good. Moreover, this method requires four evaluations per step and its global truncation error is $O\left(h^4\right)$.

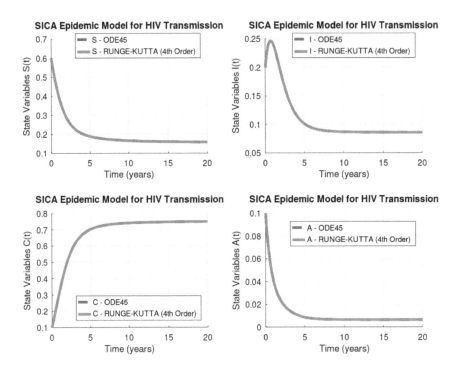

Figure 3. HIV/AIDS system (Equation (2)): GNU Octave versus Runge–Kutta's method of order four.

Table 4 lists the norm of the difference vector between results obtained by the Octave routine ode45 and the 4th order Runge–Kutta method.

Table 4. Norms 1, 2, and ∞ of the difference vector between ode45 GNU Octave and RK4 results.

System Variables	$S\left(t\right)$	$I\left(t\right)$	$C\left(t\right)$	$A\left(t\right)$
$\|Octave - RungeKutta4\|_1$	0.0003193	0.0002733	0.0004841	0.0000579
$\|Octave - RungeKutta4\|_2$	0.0000409	0.0000395	0.0000674	0.0000098
$\|Octave - RungeKutta4\|_\infty$	0.0000107	0.0000140	0.0000186	0.0000042

4. Optimal Control of HIV Transmission

In this section, we propose an optimal control problem that will be solved numerically in Octave/MATLAB in Section 4.2. We introduce a control function $u(\cdot)$ in the model of Equation (2), which represents the effort on HIV prevention measures, such as condom use (used consistently and correctly during every sex act) or oral pre-exposure prophylasis (PrEP). The control system is given by

$$\begin{cases} s'(t) = b(1 - s(t)) - (1 - u(t))\beta(i(t) + \eta_C c(t) + \eta_A a(t))s(t) + d\,a(t)\,s(t) \\ i'(t) = (1 - u(t))\beta\,(i(t) + \eta_C\,c(t) + \eta_A a(t))\,s(t) - (\rho + \phi + b)i(t) + \alpha a(t) + \omega c(t) + d\,a(t)\,i(t) \\ c'(t) = \phi i(t) - (\omega + b)c(t) + d\,a(t)\,c(t) \\ a'(t) = \rho\,i(t) - (\alpha + b + d)a(t) + d\,a^2(t), \end{cases} \tag{4}$$

where the control $u(\cdot)$ is bounded between 0 and $u_{\max}$, with $u_{\max} < 1$. When the control vanishes, no extra preventive measure for HIV transmission is being used by susceptible individuals. We assume that u_{max} is never equal to 1, since it makes the model more realistic from a medical point of view.

The goal is to find the optimal value u^* of the control u along time, such that the associated state trajectories s^*, i^*, c^*, and a^* are solutions of the system in Equation (4) in the time interval $[0, T]$ with the following initial given conditions:

$$s(0) \geq 0, \quad i(0) \geq 0, \quad c(0) \geq 0, \quad a(0) \geq 0, \tag{5}$$

and $u^*(\cdot)$ maximizes the objective functional given by

$$J(u(\cdot)) = \int_0^T \left(s(t) - i(t) - u^2(t)\right) dt, \tag{6}$$

which considers the fraction of susceptible individuals (s) and HIV-infected individuals without AIDS symptoms (i) and the cost associated with the support of HIV transmission measures (u).

The control system in Equation (4) of ordinary differential equations in $\mathbb{R}^4$ is considered with the set of admissible control functions given by

$$\Omega = \{u(\cdot) \in L^\infty(0, T) \,|\, 0 \leq u(t) \leq u_{\max}, \, \forall t \in [0, T]\}. \tag{7}$$

We consider the optimal control problem of determining $(s^*(\cdot), i^*(\cdot), c^*(\cdot), a^*(\cdot))$ associated to an admissible control $u^*(\cdot) \in \Omega$ on the time interval $[0, T]$, satisfying Equation (4) and the initial conditions of Equation (5) and maximizing the cost functional of Equation (6):

$$J(u^*(\cdot)) = \max_\Omega J(u(\cdot)). \tag{8}$$

Note that we are considering a L^2-cost function: the integrand of the cost functional J is concave with respect to the control u. Moreover, the control system of Equation (4) is Lipschitz with respect to the state variables (s, i, c, a). These properties ensure the existence of an optimal control $u^*(\cdot)$ of the optimal control problem in Equations (4)–(8) (see, e.g., Reference [14]).

To solve optimal control problems, two approaches are possible: direct and indirect. Direct methods consist in the discretization of the optimal control problem, reducing it to a nonlinear programming problem [15,16]. For such an approach, one only needs to use the Octave/MATLAB fmincon routine. Indirect methods are more sound because they are based on Pontryagin's Maximum Principle but less widespread since they are not immediately available in Octave/MATLAB. Here, we show how one can use Octave/MATLAB to solve optimal control problems through Pontryagin's Maximum Principle, reducing the optimal control problem to the solution of a boundary value problem.

4.1. Pontryagin's Maximum Principle

According to celebrated Pontryagin's Maximum Principle (see, e.g., Reference [3]), if $u^*(\cdot)$ is optimal for Equations (4)–(8) with fixed final time T, then there exists a nontrivial absolutely continuous mapping $\Lambda : [0, T] \rightarrow \mathbb{R}^4$, $\Lambda(t) = (\lambda_1(t), \lambda_2(t), \lambda_3(t), \lambda_4(t))$, called the *adjoint vector*, such that

$$s' = \frac{\partial H}{\partial \lambda_1}, \quad i' = \frac{\partial H}{\partial \lambda_2}, \quad c' = \frac{\partial H}{\partial \lambda_3}, \quad a' = \frac{\partial H}{\partial \lambda_4}, \quad \lambda_1' = -\frac{\partial H}{\partial s}, \quad \lambda_2' = -\frac{\partial H}{\partial i}, \quad \lambda_3' = -\frac{\partial H}{\partial c}, \quad \lambda_4' = -\frac{\partial H}{\partial a},$$

where

$$
\begin{aligned}
H = H\,(s(t), i(t), c(t), a(t), \Lambda(t), u(t)) = {} & s(t) - i(t) - u^2(t) \\
& + \lambda_1(t)\Big(b(1 - s(t)) - (1 - u(t))\beta(i(t) + \eta_C c(t) + \eta_A a(t))s(t) + d\,a(t)\,s(t)\Big) \\
& + \lambda_2(t)\Big((1 - u(t))\beta\,(i(t) + \eta_C\,c(t) + \eta_A a(t))\,s(t) - (\rho + \phi + b)i(t) + \alpha a(t) + \omega c(t) + d\,a(t)\,i(t)\Big) \\
& + \lambda_3(t)\Big(\phi i(t) - (\omega + b)c(t) + d\,a(t)\,c(t)\Big) \\
& + \lambda_4(t)\Big(\rho\,i(t) - (\alpha + b + d)a(t) + d\,a^2(t)\Big)
\end{aligned}
$$

is called the *Hamiltonian* and the maximality condition

$$H(s^*(t), i^*(t), c^*(t), a^*(t), \Lambda(t), u^*(t)) = \max_{0 \le u \le u_{\max}} H(s^*(t), i^*(t), c^*(t), a^*(t), \Lambda(t), u)$$

holds almost everywhere on $[0, T]$. Moreover, the transversality conditions

$$\lambda_i(T) = 0, \qquad i = 1, \dots, 4,$$

hold. Applying the Pontryagin maximum principle to the optimal control problem in Equations (4)–(8), the following theorem follows.

Theorem 1. *The optimal control problem of Equations (4)–(8) with fixed final time T admits a unique optimal solution $(s^*(\cdot), i^*(\cdot), c^*(\cdot), a^*(\cdot))$ associated to the optimal control $u^*(\cdot)$ on $[0, T]$ described by*

$$u^*(t) = \min\left\{\max\left\{0, \frac{\beta\,(i^*(t) + \eta_C c^*(t) + \eta_A a^*(t))\,s^*(t)\,(\lambda_1(t) - \lambda_2(t))}{2}\right\}, u_{\max}\right\}, \qquad (9)$$

where the adjoint functions satisfy

$$
\begin{cases}
\lambda_1'(t) = -1 + \lambda_1(t)\,(b + (1 - u^*(t))\,\beta\,(i^*(t) + \eta_C c^*(t) + \eta_A a^*(t)) - da^*(t)), \\
\qquad - \lambda_2(t)\,(1 - u^*(t))\,\beta\,(i^*(t) + \eta_C c^*(t) + \eta_A a^*(t)) \\
\lambda_2'(t) = 1 + \lambda_1^*(t)\,(1 - u^*(t))\,\beta s^*(t) - \lambda_2(t)\,((1 - u^*(t))\,\beta s^*(t) - (\rho + \phi + s^*(t)) + da^*(t)) \\
\qquad - \lambda_3(t)\phi - \lambda_4(t)\rho, \\
\lambda_3'(t) = \lambda_1(t)\,(1 - u^*(t))\,\beta \eta_C s^*(t) - \lambda_2(t)\,((1 - u^*(t))\,\beta \eta_C s^*(t) + \omega) + \lambda_3(t)\,(\omega + b - da^*(t)), \\
\lambda_4'(t) = \lambda_1(t)\,((1 - u^*(t))\,\beta \eta_A s^*(t) + ds^*(t)) - \lambda_2(t)\,((1 - u^*(t))\,\beta \eta_A s^*(t) + \alpha + di^*(t)) \\
\qquad - \lambda_3(t)dc^*(t) + \lambda_4(t)\,(\alpha + b + d - 2da^*(t)),
\end{cases}
\qquad (10)
$$

subject to the transversality conditions $\lambda_i(T) = 0$, $i = 1, \dots, 4$.

Remark 1. *The uniqueness of the optimal control u^* is due to the boundedness of the state and adjoint functions and the Lipschitz property of the systems in Equations (4) and (10) (see References [17,18] and references cited therein).*

We implement Theorem 1 numerically in Octave/MATLAB in Section 4.2, and the optimal solution $(s^*(\cdot), i^*(\cdot), c^*(\cdot), a^*(\cdot))$ associated to the optimal control $u^*(\cdot)$ is computed for given parameter values and initial conditions.

4.2. Numerical Solution of the HIV Optimal Control Problem

The extremal given by Theorem 1 is now computed numerically by implementing a forward-backward fourth-order Runge–Kutta method (see, e.g., Reference [19]). This iterative method consists in solving the system in Equation (4) with a guess for the controls over the time interval $[0, T]$ using a forward fourth-order Runge–Kutta scheme and the transversality conditions $\lambda_i(T) = 0, i = 1, \ldots, 4$. Then, the adjoint system in Equation (10) is solved by a backward fourth-order Runge–Kutta scheme using the current iteration solution of Equation (4). The controls are updated by using a convex combination of the previous controls and the values from Equation (9). The iteration is stopped if the values of unknowns at the previous iteration are very close to the ones at the present iteration. Our odeRungeKutta_order4_WithControl function is implemented by the following GNU Octave instructions:

```
function dy = odeRungeKutta_order4_WithControl(T)
  % Parameters of the model
  mi = 1.0 / 69.54; b = 2.1 * mi; beta = 1.6;
  etaC = 0.015; etaA = 1.3; fi = 1.0; ro = 0.1;
  alfa = 0.33; omega = 0.09; d = 1.0;

  % Parameters of the Runge-Kutta (4th order) method
  test = -1; deltaError = 0.001; M = 1000;
  t = linspace(0,T,M+1);
  h = T / M; h2 = h / 2; h6 = h / 6;
  S = zeros(1,M+1); I = zeros(1,M+1);
  C = zeros(1,M+1); A = zeros(1,M+1);

  % Initial conditions of the model
  S(1) = 0.6; I(1) = 0.2; C(1) = 0.1; A(1) = 0.1;

  %Vectors for system restrictions and control
  Lambda1 = zeros(1,M+1); Lambda2 = zeros(1,M+1);
  Lambda3 = zeros(1,M+1); Lambda4 = zeros(1,M+1);
  U = zeros(1,M+1);
  % Iterations of the method
  while(test < 0)
    oldS = S; oldI = I; oldC = C; oldA = A;
    oldLambda1 = Lambda1; oldLambda2 = Lambda2;
    oldLambda3 = Lambda3; oldLambda4 = Lambda4;
    oldU = U;

    %Forward Runge-Kutta iterations
    for i = 1:M
      % Differential equations of the model
      % First Runge-Kutta parameter
      aux1 = (1 - U(i)) * beta * (I(i) + etaC * C(i) + etaA * A(i)) * S(i);
      aux2 = d * A(i);

      auxS1 = b * (1 - S(i)) - aux1 + aux2 * S(i);
      auxI1 = aux1 - (ro + fi + b - aux2) * I(i) + alfa * A(i) + omega * C(i);
      auxC1 = fi * I(i) - (omega + b - aux2) * C(i);
      auxA1 = ro * I(i) - (alfa + b + d - aux2) * A(i);
```

```
    % Second Runge-Kutta parameter
    auxU = 0.5 * (U(i) + U(i+1));
    auxS = S(i) + h2 * auxS1; auxI = I(i) + h2 * auxI1;
    auxC = C(i) + h2 * auxC1; auxA = A(i) + h2 * auxA1;
    aux1 = (1 - auxU) * beta * (auxI + etaC * auxC + etaA * auxA) * auxS;
    aux2 = d * auxA;
    auxS2 = b * (1 - auxS) - aux1 + aux2 * auxS;
    auxI2 = aux1 - (ro + fi + b - aux2) * auxI + alfa * auxA + omega * auxC;
    auxC2 = fi * auxI - (omega + b - aux2) * auxC;
    auxA2 = ro * auxI - (alfa + b + d - aux2) * auxA;


    % Third Runge-Kutta parameter
    auxS = S(i) + h2 * auxS2; auxI = I(i) + h2 * auxI2;
    auxC = C(i) + h2 * auxC2; auxA = A(i) + h2 * auxA2;
    aux1 = (1 - auxU) * beta * (auxI + etaC * auxC + etaA * auxA) * auxS;
    aux2 = d * auxA;

    auxS3 = b * (1 - auxS) - aux1 + aux2 * auxS;
    auxI3 = aux1 - (ro + fi + b - aux2) * auxI + alfa * auxA + omega * auxC;
    auxC3 = fi * auxI - (omega + b - aux2) * auxC;
    auxA3 = ro * auxI - (alfa + b + d - aux2) * auxA;


    % Fourth Runge-Kutta parameter
    auxS = S(i) + h * auxS3; auxI = I(i) + h * auxI3;
    auxC = C(i) + h * auxC3; auxA = A(i) + h * auxA3;
    aux1 = (1 - U(i+1)) * beta * (auxI + etaC * auxC + etaA * auxA) * auxS;
    aux2 = d * auxA;

    auxS4 = b * (1 - auxS) - aux1 + aux2 * auxS;
    auxI4 = aux1 - (ro + fi + b - aux2) * auxI + alfa * auxA + omega * auxC;
    auxC4 = fi * auxI - (omega + b - aux2) * auxC;
    auxA4 = ro * auxI - (alfa + b + d - aux2) * auxA;


    % Runge-Kutta new approximation
    S(i+1) = S(i) + h6 * (auxS1 + 2 * (auxS2 + auxS3) + auxS4);
    I(i+1) = I(i) + h6 * (auxI1 + 2 * (auxI2 + auxI3) + auxI4);
    C(i+1) = C(i) + h6 * (auxC1 + 2 * (auxC2 + auxC3) + auxC4);
    A(i+1) = A(i) + h6 * (auxA1 + 2 * (auxA2 + auxA3) + auxA4);
end

%Backward Runge-Kutta iterations
for i = 1:M
  j = M + 2 - i;

  % Differential equations of the model
  % First Runge-Kutta parameter
  auxU = 1 - U(j);
  aux1 = auxU * beta * (I(j) + etaC * C(j) + etaA * A(j));
  aux2 = d * A(j);

  auxLambda11 = -1 + Lambda1(j) * (b + aux1 - aux2) - Lambda2(j) * aux1;
  aux1 = auxU * beta * S(j);
  auxLambda21 = 1 + Lambda1(j) * aux1 - Lambda2(j) * (aux1 - (ro + fi + b) + ...
                + aux2) - Lambda3(j) * fi - Lambda4(j) * ro;
```

```
aux1 = auxU * beta * etaC * S(j);
auxLambda31 = Lambda1(j) * aux1 - Lambda2(j) * (aux1 + ...
              + omega) + Lambda3(j) * (omega + b - aux2);
aux1 = auxU * beta * etaA * S(j);
auxLambda41 = Lambda1(j) * (aux1 + d * S(j)) ...
              - Lambda2(j) * (aux1 + alfa + ...
              + d * I(j)) - Lambda3(j) * d * C(j) + ...
              + Lambda4(j) * (alfa + b + d - 2 * aux2);


% Second Runge-Kutta parameter
auxU = 1 - 0.5 * (U(j) + U(j-1));
auxS = 0.5 * (S(j) + S(j-1));
auxI = 0.5 * (I(j) + I(j-1));
auxC = 0.5 * (C(j) + C(j-1));
auxA = 0.5 * (A(j) + A(j-1));


aux1 = auxU * beta * (auxI + etaC * auxC + etaA * auxA);
aux2 = d * auxA;
auxLambda1 = Lambda1(j) - h2 * auxLambda11;
auxLambda2 = Lambda2(j) - h2 * auxLambda21;
auxLambda3 = Lambda3(j) - h2 * auxLambda31;
auxLambda4 = Lambda4(j) - h2 * auxLambda41;


auxLambda12 = -1 + auxLambda1 * (b + aux1 - aux2) - auxLambda2 * aux1;
aux1 = auxU * beta * auxS;
auxLambda22 = 1 + auxLambda1 * aux1 - auxLambda2 * (aux1 - (ro + fi + b) + ...
              + aux2) - auxLambda3 * fi - auxLambda4 * ro;
aux1 = auxU * beta * etaC * auxS;
auxLambda32 = auxLambda1 * aux1 - auxLambda2 * (aux1 + ...
              + omega) + auxLambda3 * (omega + b - aux2);
aux1 = auxU * beta * etaA * auxS;
auxLambda42 = auxLambda1 * (aux1 + d * auxS) ...
              - auxLambda2 * (aux1 + alfa + ...
              + d * auxI) - auxLambda3 * d * auxC + ...
              + auxLambda4 * (alfa + b + d - 2 * aux2);


% Third Runge-Kutta parameter
aux1 = auxU * beta * (auxI + etaC * auxC + etaA * auxA);
auxLambda1 = Lambda1(j) - h2 * auxLambda12;
auxLambda2 = Lambda2(j) - h2 * auxLambda22;
auxLambda3 = Lambda3(j) - h2 * auxLambda32;
auxLambda4 = Lambda4(j) - h2 * auxLambda42;


auxLambda13 = -1 + auxLambda1 * (b + aux1 - aux2) - auxLambda2 * aux1;
aux1 = auxU * beta * auxS;
auxLambda23 = 1 + auxLambda1 * aux1 ...
              - auxLambda2 * (aux1 - (ro + fi + b) + ...
              + aux2) - auxLambda3 * fi - auxLambda4 * ro;
aux1 = auxU * beta * etaC * auxS;
auxLambda33 = auxLambda1 * aux1 - auxLambda2 * (aux1 + ...
              + omega) + auxLambda3 * (omega + b - aux2);
aux1 = auxU * beta * etaA * auxS;
auxLambda43 = auxLambda1 * (aux1 + d * auxS) ...
              - auxLambda2 * (aux1 + alfa + d * auxI) - auxLambda3 * d * auxC + ...
              + auxLambda4 * (alfa + b + d - 2 * aux2);
```

```
% Fourth Runge-Kutta parameter
auxU = 1 - U(j-1); auxS = S(j-1);
auxI = I(j-1); auxC = C(j-1); auxA = A(j-1);


aux1 = auxU * beta * (auxI + etaC * auxC + etaA * auxA);
aux2 = d * auxA;
auxLambda1 = Lambda1(j) - h * auxLambda13;
auxLambda2 = Lambda2(j) - h * auxLambda23;
auxLambda3 = Lambda3(j) - h * auxLambda33;
auxLambda4 = Lambda4(j) - h * auxLambda43;


auxLambda14 = -1 + auxLambda1 * (b + aux1 - aux2) ...
              - auxLambda2 * aux1;
aux1 = auxU * beta * auxS;
auxLambda24 = 1 + auxLambda1 * aux1 ...
              - auxLambda2 * (aux1 - (ro + fi + b) + ...
              + aux2) - auxLambda3 * fi - auxLambda4 * ro;
aux1 = auxU * beta * etaC * auxS;
auxLambda34 = auxLambda1 * aux1 - auxLambda2 * (aux1 + ...
              + omega) + auxLambda3 * (omega + b - aux2);
aux1 = auxU * beta * etaA * auxS;
auxLambda44 = auxLambda1 * (aux1 + d * auxS) ...
              - auxLambda2 * (aux1 + alfa + ...
              + d * auxI) - auxLambda3 * d * auxC + ...
              + auxLambda4 * (alfa + b + d - 2 * aux2);


% Runge-Kutta new approximation
Lambda1(j-1) = Lambda1(j) - h6 * (auxLambda11 + ...
               + 2 * (auxLambda12 + auxLambda13) + auxLambda14);
Lambda2(j-1) = Lambda2(j) - h6 * (auxLambda21 + ...
               + 2 * (auxLambda22 + auxLambda23) + auxLambda24);
Lambda3(j-1) = Lambda3(j) - h6 * (auxLambda31 + ...
               + 2 * (auxLambda32 + auxLambda33) + auxLambda34);
Lambda4(j-1) = Lambda4(j) - h6 * (auxLambda41 + ...
               + 2 * (auxLambda42 + auxLambda43) + auxLambda44);
end

% New vector control
for i = 1:M+1
  vAux(i) = 0.5 * beta * (I(i) + etaC * C(i) + ...
            + etaA * A(i)) * S(i) * (Lambda1(i) - Lambda2(i));
  auxU = min([max([0.0 vAux(i)]) 0.5]);
  U(i) = 0.5 * (auxU + oldU(i));
end

% Absolute error for convergence
temp1 = deltaError * sum(abs(S)) - sum(abs(oldS - S));
temp2 = deltaError * sum(abs(I)) - sum(abs(oldI - I));
temp3 = deltaError * sum(abs(C)) - sum(abs(oldC - C));
temp4 = deltaError * sum(abs(A)) - sum(abs(oldA - A));
temp5 = deltaError * sum(abs(U)) - sum(abs(oldU - U));
temp6 = deltaError * sum(abs(Lambda1)) - sum(abs(oldLambda1 - Lambda1));
temp7 = deltaError * sum(abs(Lambda2)) - sum(abs(oldLambda2 - Lambda2));
temp8 = deltaError * sum(abs(Lambda3)) - sum(abs(oldLambda3 - Lambda3));
```

```
        temp9 = deltaError * sum(abs(Lambda4)) - sum(abs(oldLambda4 - Lambda4));
    test = min(temp1,min(temp2,min(temp3,min(temp4, ...
            min(temp5,min(temp6,min(temp7,min(temp8,temp9)))))))));
end
dy(1,:)  = t; dy(2,:)  = S; dy(3,:)  = I;
dy(4,:)  = C; dy(5,:)  = A; dy(6,:)  = U;

disp("Value of LAMBDA at FINAL TIME");
disp([Lambda1(M+1) Lambda2(M+1) Lambda3(M+1) Lambda4(M+1)]);
```

For the numerical simulations, we consider $u_{max} = 0.5$, representing a lack of resources or misuse of the preventive HIV measures $u(\cdot)$, that is, the set of admissible controls is given by

$$\Omega = \{u(\cdot) \in L^{\infty}(0, T) \,|\, 0 \le u(t) \le 0.5 \,, \forall t \in [0, T]\} \tag{11}$$

with $T = 20$ (years). Figures 4 and 5 show the numerical solution to the optimal control problem of Equations (4)–(8) with the initial conditions of Equation (3) and the admissible control set in Equation (11) computed by our odeRungeKutta_order4_WithControl function. Figure 6 depicts the extremal control behaviour of u^*.

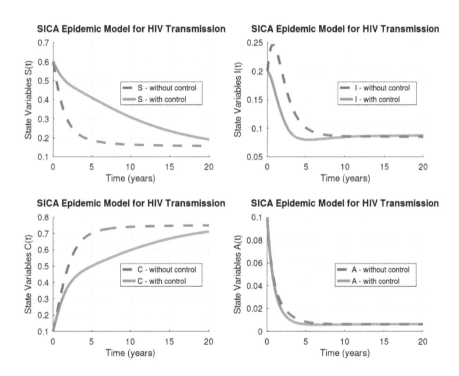

Figure 4. Optimal state variables for the control problem in Equations (4)–(8) subject to the initial conditions in Equation (3) and the admissible control set in Equation (11) *versus* trajectories without control measures.

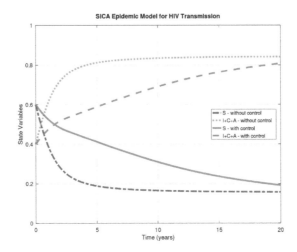

Figure 5. Comparison: solutions to the initial value problem in Equations (2)–(3) *versus* solutions to the optimal control problem in Equations (4)–(8) subject to initial conditions in Equation (3) and the admissible control set in Equation (11).

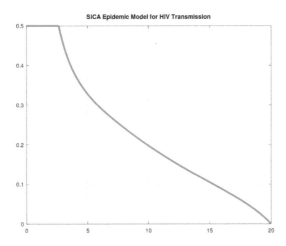

Figure 6. Optimal control u^* for the HIV optimal control problem in Equations (4)–(8) subject to the initial conditions in Equation (3) and the admissible control set in Equation (11).

5. Conclusions

The paper provides a study on numerical methods to deal with modelling and optimal control of epidemic problems. Simple but effective Octave/MATLAB code is fully provided for a recent model proposed in Reference [10]. The given numerical procedures are robust with respect to the parameters: we have used the same values as the ones in Reference [10], but the code is valid for other values of the parameters and easily modified to other models. The results show the effectiveness of optimal control theory in medicine and the usefulness of a scientific computing system such as GNU Octave: using the control measure as predicted by Pontryagin's maximum principle and numerically computed by our Octave code, one sees that the number of HIV/AIDS-infected and -chronic individuals diminish and, as a consequence, the number of susceptible (not ill) individuals increase. We trust our paper will be very useful to a practitioner from the disease control area. Indeed, this work has been motivated by many emails we received and continue to receive, asking us to provide the software code associated to our research papers on applications of optimal control theory in epidemiology, e.g., References [20,21].

Author Contributions: Each author equally contributed to this paper, read and approved the final manuscript. All authors have read and agreed to the published version of the manuscript.

Funding: This research was partially supported by the Portuguese Foundation for Science and Technology (FCT) within projects UID/MAT/04106/2019 (CIDMA) and PTDC/EEI-AUT/2933/2014 (TOCCATTA) and was cofunded by FEDER funds through COMPETE2020—Programa Operacional Competitividade e Internacionalização (POCI) and by national funds (FCT). Silva is also supported by national funds (OE), through FCT, I.P., in the scope of the framework contract foreseen in numbers 4–6 of article 23 of the Decree-Law 57/2016 of August 29, changed by Law 57/2017 of July 19.

Acknowledgments: The authors are sincerely grateful to four anonymous reviewers for several useful comments, suggestions, and criticisms, which helped them to improve the paper.

Conflicts of Interest: The authors declare no conflict of interest.

References

1. Brauer, F.; Castillo-Chavez, C. *Mathematical Models in Population Biology and Epidemiology*, 2nd ed.; Springer: New York, NY, USA, 2012. [CrossRef]

2. Elazzouzi, A.; Lamrani Alaoui, A.; Tilioua, M.; Torres, D.F.M. Analysis of a SIRI epidemic model with distributed delay and relapse. *Stat. Optim. Inf. Comput.* **2019**, *7*, 545–557. [CrossRef]

3. Pontryagin, L.S.; Boltyanskii, V.G.; Gamkrelidze, R.V.; Mishchenko, E.F. *The Mathematical Theory of Optimal Processes*; Translated from the Russian by K.N. Trirogoff; Neustadt, L.W., Ed.; John Wiley & Sons, Inc.: New York, NY, USA, 1962.

4. Area, I.; Ndaïrou, F.; Nieto, J.J.; Silva, C.J.; Torres, D.F.M. Ebola model and optimal control with vaccination constraints. *J. Ind. Manag. Optim.* **2018**, *14*, 427–446. [CrossRef]

5. Burlacu, R.; Cavache, A. On a class of optimal control problems in mathematical biology. *IFAC Proc. Vol.* **1999**, *32*, 3746–3759. [CrossRef]

6. Deshpande, S. *Optimal Input Signal Design for Data-Centric Identification and Control with Applications to Behavioral Health and Medicine*; ProQuest LLC: Ann Arbor, MI, USA, 2014.

7. Silva, C. J.; Torres, D. F. M.; Venturino, E. Optimal spraying in biological control of pests. *Math. Model. Nat. Phenom.* **2017**, *12*, 51–64. [CrossRef]

8. Rosa, S.; Torres, D.F.M. Optimal control and sensitivity analysis of a fractional order TB model. *Stat. Optim. Inf. Comput.* **2019**, *7*, 617–625. [CrossRef]

9. Allali, K.; Harroudi, S.; Torres, D.F.M. Analysis and optimal control of an intracellular delayed HIV model with CTL immune response. *Math. Comput. Sci.* **2018**, *12*, 111–127. [CrossRef]

10. Silva, C.J.; Torres, D.F.M. A SICA compartmental model in epidemiology with application to HIV/AIDS in Cape Verde. *Ecol. Complex.* **2017**, *30*, 70–75. [CrossRef]

11. Silva, C.J.; Torres, D.F.M. Modeling and optimal control of HIV/AIDS prevention through PrEP. *Discrete Contin. Dyn. Syst. Ser. S* **2018**, *11*, 119–141. [CrossRef]

12. An, G. The Crisis of Reproducibility, the Denominator Problem and the Scientific Role of Multi-scale Modeling. *Bull. Math. Biol.* **2018**, *80*, 3071–3080. [CrossRef] [PubMed]

13. Eaton, J.W.; Bateman, D.; Hauberg, S.; Wehbring, R. GNU Octave Version 5.1.0 Manual: A hIgh-Level Interactive Language for Numerical Computations. Available online: https://enacit.epfl.ch/cours/matlab-octave/octave-documentation/octave/octave.pdf (accessed on 6 October 2019).

14. Cesari, L. *Optimization—Theory and Applications*; Springer: New York, NY, USA, 1983. [CrossRef]

15. Salati, A.B.; Shamsi, M.; Torres, D.F.M. Direct transcription methods based on fractional integral approximation formulas for solving nonlinear fractional optimal control problems. *Commun. Nonlinear Sci. Numer. Simul.* **2019**, *67*, 334–350. [CrossRef]

16. Nemati, S.; Lima, P.M.; Torres, D.F.M. A numerical approach for solving fractional optimal control problems using modified hat functions. *Commun. Nonlinear Sci. Numer. Simul.* **2019**, *78*, 104849. [CrossRef]

Math. Comput. Appl. **2020**, *25*, 1

17. Jung, E.; Lenhart, S.; Feng, Z. Optimal control of treatments in a two-strain tuberculosis model. *Discrete Contin. Dyn. Syst. Ser. B* **2002**, *2*, 473–482. [CrossRef]

18. Silva, C.J.; Torres, D.F.M. Optimal control for a tuberculosis model with reinfection and post-exposure interventions. *Math. Biosci.* **2013**, *244*, 154–164. [CrossRef] [PubMed]

19. Lenhart, S.; Workman, J.T. *Optimal Control Applied to Biological Models*; Chapman & Hall/CRC: Boca Raton, FL, USA, 2007.

20. Rachah, A.; Torres, D.F.M. Mathematical modelling, simulation, and optimal control of the 2014 Ebola outbreak in West Africa. *Discrete Dyn. Nat. Soc.* **2015**. [CrossRef]

21. Malinzi, J.; Ouifki, R.; Eladdadi, A.; Torres, D.F.M.; White, K.A.J. Enhancement of chemotherapy using oncolytic virotherapy: Mathematical and optimal control analysis. *Math. Biosci. Eng.* **2018**, *15*, 1435–1463. [CrossRef] [PubMed]

Mathematical and Computational Applications

Article

Isogeometric Analysis for Fluid Shear Stress in Cancer Cells

José A. Rodrigues

Centro de Investigação em Modelação e Optimização de Sistemas Multifuncionais (CIMOSM), Instituto Superior de Engenharia de Lisboa (ISEL), Instituto Politécnico de Lisboa, Rua Conselheiro Emídio Navarro, 1959-007 Lisboa, Portugal; jose.rodrigues@isel.pt

Received: 20 December 2019; Accepted: 2 April 2020; Published: 3 April 2020

Abstract: The microenvironment of the tumor is a key factor regulating tumor cell invasion and metastasis. The effects of physical factors in tumorigenesis is unclear. Shear stress, induced by liquid flow, plays a key role in proliferation, apoptosis, invasion, and metastasis of tumor cells. The mathematical models have the potential to elucidate the metastatic behavior of the cells' membrane exposed to these microenvironment forces. Due to the shape configuration of the cancer cells, Non-uniform Rational B-splines (NURBS) lines are very adequate to define its geometric model. The Isogeometric Analysis allows a simplified transition of exact CAD models into the analysis avoiding the geometrical discontinuities of the traditional Galerkin traditional techniques. In this work, we use an isogeometric analysis to model the fluid-generated forces that tumor cells are exposed to in the vascular and tumor microenvironments, in the metastatic process. Using information provided by experimental tests in vitro, we present a suite of numerical experiments which indicate, for standard configurations, the metastatic behavior of cells exposed to such forces. The focus of this paper is strictly on geometrical sensitivities to the shear stress' exhibition for the cell membrane, this being its innovation.

Keywords: Darcy; Brinkman; incompressible; isogeometric analysis; shear stress; interstitial flow; cancer; NURBS

---

1. Introduction

The formation of a secondary tumor at a site distant from the the primary tumor is known as metastasis by a cancerous tumor. To initiate the metastatic spread of cancer through the bloodstream, tumor cells must transit through microenvironments of dramatically varying physical forces. Cancer cells are able to migrate through the stroma, intravasate through the endothelium into the blood or lymphatic vessels, to flow within the vessels, to extravasate from the vessel through the endothelium and colonize in tissue at a secondary site [1].

In soft tissues, cancer cells are exposed to mechanical forces due to fluid shear stress, hydrostatic pressure, tension and compression forces. Fluid shear stress is one of the most important forces that cells are exposed to, and its effects on blood cells, endothelial cells, smooth muscle cells, and others have been extensively studied. However, much less is known about fluid shear stress effects on tumor cells. Cancer cells experience two main kinds of fluid shear stress: stresses generated by blood flow in the vascular microenvironment, and those generated by interstitial flows in the tumor microenvironment [2]. Stresses generated by interstitial and blood flows could contribute to the metastatic process by enhancing tumor cell invasion and circulating tumor cell adhesion to blood vessels, respectively. However, it is difficult to predict tumor cell behavior to such forces and it is difficult to experimentally measure such flows in the tumor microenvironment [3].

In this work, we apply Mathematical and Mechanical processes to analyze the metastatic behavior of the cells' membrane exposed to microenvironement forces. Providing an Isogeometric Analysis

(IGA) [4] computational method to model and predict how cancer cells respond to such forces, we allow for new insight and new decision tools for medical problems.

The biological fluid study at microenvironments, in vivo or in vitro, is a recent topic following the microfluidic technology and mechanical methodology methods developments [5–8], with the ethical and budget restrictions, increasing the evidence that fluid shear stress is an essential factor affecting fluid mechanics [1].

The current work serves as an introduction to this line of research on mathematical modeling to help us understand the omics data produced by experimental techniques and to bridge the gap between the developments of technologies and systemic modelling of the biological process in cancer research.

Although it is accepted that the effectiveness of IGA methods is well-established for problems with complex geometries, its effect in biologic models has not been extensively studied. Indeed, at present, general research in this subject is still in this infancy for the case studies presented [9,10].

First, we introduce and describe the methodology. We then provide a mathematical and numerical analysis overview for the interstitial flow governing model and in order to estimate the fluid shear stress on cells in tissues.

Standard shape for cancer cells is very irregular, for this reason Non-uniform Rational B-splines (NURBS) lines are very adequate to define its geometric model. The Isogeometric Analysis allows for simplified transition of exact CAD models into the analysis, avoiding the geometrical discontinuities of the traditional Galerkin traditional techniques. For the exhibit models, NURBS provides a high-quality geometric model, quite analogous to a physical model, and also defines the basis of the discrete space in which the partial differential equation solution is approximated with great accuracy per degree of freedom, as referred to in Section 3.

We conclude with simulations to predict the effect of fluid shear stress on cancer cells for several scenarios of microenvironment tumor implantation.

In this paper, we focus on the effects of shear stress, induced by liquid flow, on apoptosis, invasion, and metastasis of tumor cells, following the theoretical [11] and numerical [12] work for a Stokes flow.

2. Materials and Methods

2.1. Model Geometry

Tumor cells have a large variety in shape and size. Therefore, several different geometries will be considered in this study, always considering these as standard cells. We will consider bidimensional projections (as shown in Figure 24) of the model shown in Figure 1 to inspire the geometrical model for the presented cases.

Figure 1. Tridimensional tumor's geometrical model [7]. Bidimensional projections of this model will be considered for the study cases.

For these geometrical models, Non-uniform Rational B-splines (NURBS) lines are very adequate to define it due to simplified transition of exact CAD models into the analysis, avoiding the geometrical discontinuities of the traditional Galerkin traditional techniques.

2.2. Governing Equations

Average equations describing viscous flow through a porous region are of great theoretical and practical interest. At the microscale level, the Stokes equations apply and provide a complete description of the entire flow field. However, Darcy's law, formally derived by performing appropriate volume averages of Stokes equations, is applicable. The qualitative difference between these two flow descriptions motivate Brinkman to suggest a general equation that interpolates between the Stokes equation and Darcy's law [13]. We introduce and describe the mathematical model in the next section.

3. Mathematical Formulation

3.1. The Darcy–Brinkman Equation

Darcy's law is a phenomenologically derived constitutive equation that describes the flow of a fluid through a porous medium [9,10]. This law, as an expression of conservation of momentum, was determined experimentally by Darcy. It has since been derived from the Navier–Stokes equations via homogenization. It is analogous to Fourier's law in the field of heat conduction.

$$\vec{u} = -\frac{k}{\nu}\nabla p \tag{1}$$

where k is the permeability of the medium, ∇p is the pressure gradient vector, ν is the viscosity of the fluid and $\vec{u}$ is the fluid's average velocity through a plane region represented by Ω. Region Ω is supposed regular enough to ensure the later theoretical regularity for $\vec{u}$ and p.

To account for interstitial flows between boundaries, Brinkman has developed a second-order term, taking into account no-slip boundary conditions cells' membrane (Figure 2).

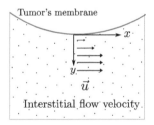

Figure 2. Outline of cells membrane interstitial zone and no-slip boundary conditions representation.

For governing interstitial flow between boundaries we will consider the Darcy–Brinkman equation (2)

$$-\nu\Delta\vec{u} + \frac{\nu}{k}\vec{u} + \nabla p = 0 \tag{2}$$

The mass balance equation for a steady state incompressible fluid is that the divergence of the fluid is zero

$$\nabla \cdot \vec{u} = 0 \tag{3}$$

To apply the boundary conditions we decompose the boundary of the region Ω in two non-overlapping regions, as shown in Figure 3, $\partial\Omega = \Gamma_T \cup \Gamma_I$, the boundary of the tumor region and the boundary of the interstitial region considered, respectively. We consider homogeneous Dirichlet boundary condition for the velocity over the cells membrane and non-homogeneous on the interstitial region boundaries, i.e., $\vec{u} = (u_x^0, u_y^0)$ and parameters ν and k given by Table 1, for the cases under study.

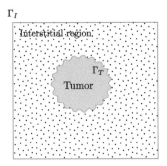

Figure 3. Outline of boundary decomposition $\partial \Omega = \Gamma_T \cup \Gamma_I$.

To facilitate understanding, we write the component-wise formulation in two dimensions. Let $\vec{u} = (u_x, u_y)$. Then Equations (2) and (3) consists of three equations

$$
\begin{aligned}
-\nu \Delta u_x + \tfrac{\nu}{k} u_x + \tfrac{\partial p}{\partial x} &= 0 \\
-\nu \Delta u_y + \tfrac{\nu}{k} u_y + \tfrac{\partial p}{\partial y} &= 0 \\
\tfrac{\partial u_x}{\partial y} + \tfrac{\partial u_y}{\partial y} &= 0
\end{aligned}
\tag{4}
$$

We are interested to find $\vec{u} = (u_x, u_y)$ and p solution of the problem defined by Equation (4) and the Dirichlet boundary conditions showed at Table 1.

With the solution, $\vec{u} = (u_x, u_y)$ we evaluate the fluid shear stress. For bidimensional Newtonian fluid the Stress is written [14]

$$
T = - \begin{bmatrix} p & 0 \\ 0 & p \end{bmatrix} + 2\nu \begin{bmatrix} \frac{\partial u_x}{\partial x} & \frac{1}{2}\left(\frac{\partial u_x}{\partial y} + \frac{\partial u_y}{\partial x}\right) \\ \frac{1}{2}\left(\frac{\partial u_y}{\partial x} + \frac{\partial u_x}{\partial y}\right) & \frac{\partial u_y}{\partial y} \end{bmatrix}
\tag{5}
$$

and the Shear Stress is described by

$$
T_{xy} = \nu \frac{1}{2} \left(\frac{\partial u_x}{\partial y} + \frac{\partial u_y}{\partial x} \right).
\tag{6}
$$

3.2. Isogeometric Discretization

In order to represent complex shapes, the use of polynomials or rational segments may often be inadequate or imprecise. On the other hand, B-spline and NURBS functions enjoy some major advantages that make them extremely convenient for complex geometrical representations.

The main idea behind the isogeometric approach [4] is to discretize the unknowns of the problem with the same set of basis functions that CAD employs for the construction of geometries. Let p be the prescribed degree and n control points, we define by

$$
\Xi = \left\{ t_1, \cdots, t_{n+p+1} \right\}
\tag{7}
$$

the knots vector, with $t_i \in [0,1]$, $i \in \{1, \cdots, n+p+1\}$. Cox-De Boor's formula [15] defines n one-dimensional B-spline basis functions recursively as

$$
B_{i,0}(t) = \begin{cases} 1 & t_i \leq t < t_{i+1} \\ 0 & \text{otherwise} \end{cases}
\tag{8}
$$

$$
B_{i,p}(t) = \frac{t - t_i}{t_{i+p} - t_i} B_{i,p-1}(t) + \frac{t_{i+p+1} - t}{t_{i+p+1} - t_{i+1}} B_{i+1,p-1}(t)
$$

for $i \in \{1, \cdots, n\}$

As referred to in [12] the support of a B-spline of degree p is always $p + 1$ knot spans and, as a consequence, each p-th degree function has $p - 1$ continuous derivatives across the element boundaries, or across the knots, if they are not repeated. Repetition of knots can be exploited to prescribe the regularity.

NURBS of degree p are defined as rational B-splines, associating to each B-spline function a weight w_i

$$N_{i,p}(t) = \frac{w_i B_{i,p}(t)}{\sum_j w_j B_{j,p}(t)}, \tag{9}$$

with w_i called the weight parameter. Geometries in the projective space may be described by using the concept of homogeneous coordinates, which are frequently denoted as weights w_i. A weighted polynomial B-spline geometry of $\mathbb{R}^{d+1}$ is obtained by first multiplying its control point data with the homogeneous coordinates. For values $w_i > 1$, the object moves toward the control polygon, whereas for weights smaller than one, the influence of the control point on the geometry decreases. Control points with $w_i = 0$ do not affect the geometric object at all. If all w_i are equal to one, the NURBS basis simplifies to the polynomial B-spline basis [15] and allows for the partition of unity property.

As in [12] we define bidimensional B-splines and NURBS using a tensor product approach. Considering $\Xi = \Xi_1 \times \Xi_2$ the knot vectors, $\mathbf{p}$ the degrees and $\mathbf{n}$ the number of basis functions the bivariate B-spline is given by

$$\mathbf{N}_{\mathbf{i},\mathbf{p}}(\mathbf{t}) = N_{i_1,p_1}(t_1) N_{i_2,p_2}(t_2) \tag{10}$$

where $\mathbf{t} = (t_1, t_2)$, $\mathbf{p} = (p_1, p_2)$, $\mathbf{n} = (n_1, n_2)$ and $\mathbf{i} = (i_1, i_2)$ is a multi-index in the set $i_1 \in \{1, \cdots, n_1\}$ and $i_2 \in \{1, \cdots, n_2\}$.

It is straightforward to notice that there is a parametric Cartesian mesh $\mathcal{T}_h$ associated with Ξ. The knot vectors partitioning the parametric domain $[0, 1]^2$ into parallelograms. For each element $Q \in \mathcal{T}_h$ we associate a parametric mesh size $h_Q = \text{diam}(Q)$, and $h = \max \{h_Q, Q \in \mathcal{T}_h\}$.

In the following, we refer to the basis functions indicating the global index, and we will denote by $S_h(\Xi)$ the bivariate B-spline space spanned by the basis functions $\mathbf{N_i}$, $1 \le i \le n$. For convenience we also use the notation $S_{\alpha_1,\alpha_2}^{p_1,p_2}(\Xi)$ to designate the associated space of splines of order p_1 in the x direction, p_2 in the y direction, and smoothness α_1 and α_2 respectively. Notice that α_i is determined by the knot vector Ξ_i, spanned by the basis functions of degree p_d and regularity α_d for each direction $d \in \{1, 2\}$.

A NURBS curve is defined by a set of control points $\mathbf{P}$ which act as weights for the linear combination of the basis functions, giving the mapping to the physical space. In particular, given n one-dimensional basis functions $N_{i,p}$ and n control points $\mathbf{P}_i \in \mathbb{R}^2$, $i \in \{1, \cdots, n\}$ a curve parameterization is given by:

$$\mathbf{C}(t) = \sum_{i=1}^{n} \mathbf{P}_i N_i(t). \tag{11}$$

The control points define the so-called control mesh but this does not, in general, conform to the actual geometry (cf Figure 5b). In many real-world applications, the computational domain may be too complicated to be represented by a single NURBS mapping from the reference domain to the physical space. This could be due to topological reasons (or to the presence of different materials). In these cases it is common practice to resort to the so-called multipatch approach [4,12] (cf Figure 4). Here, the physical domain is split into simpler subdomains Ω_k such as $\Omega = \cup_k \Omega_k$ and $\Omega_i \cap \Omega_j = \emptyset$, or a point, or an edge, for $i \neq j$.

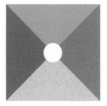

Figure 4. Case 1: the multipatch domain.

3.3. Isogeometric Conforming Spaces

Analogously to classical Finite Element Methods, IGA is based on a Galerkin approach: the equations are written in their variational formulation, and the solution is sought in a finite-dimensional space with the correct approximation properties. In IGA the basis function space is inherited from the one used to parametrize the geometry.

Let us now consider a domain $\Omega \subset V$ that can be exactly parametrized with a mapping $\vec{F}$

$$\vec{F} : [0,1]^2 \to \Omega \tag{12}$$

For a multipatch approach we will consider a map for each Ω_k.

Then, the discrete space in the physical domain is defined applying the isoparametric concept as

$$Q_h \;=\; \left\{ q_h := \eta \circ \vec{F}^{-1}, \quad \eta \in S^{p_1,p_2}_{\alpha_1,\alpha_2}(\Xi) \right\} \tag{13}$$

$$\mathbf{V}_h \;=\; \left\{ \vec{v}_h := \vec{\varphi} \circ \vec{F}^{-1}, \quad \vec{\varphi} \in \left(S^{p_1+1,p_2+1}_{\alpha_1,\alpha_2}(\Pi) \right)^2 \right\}, \tag{14}$$

where each Π has the same knots Ξ but the multiplicity has been increased by one, that means the velocity components' space have the same continuity as the pressure space [16]. We recall the important property of B-spline: at a knot of multiplicity m, basis function $N_{i,p}$ is $C^{p-m} = C^\alpha$ continuous.

Proposition 1. *With the notation and assumptions above, the space*

$$M_h = \{ q_h = \nabla \cdot \vec{v}_h, \ \vec{v}_h \in \mathbf{V_h} \}. \tag{15}$$

is subspace of Q_h.

Proof of Proposition 1. For $\vec{w} \in \left(S^{p_1+1,p_2+1}_{\alpha_1,\alpha_2}(\Pi) \right)^2$ we obtain

$$\nabla \cdot \vec{w} \in S^{p_1,p_2}_{\alpha_1-1,\alpha_2-1}(\Pi) = S^{p_1,p_2}_{\alpha_1,\alpha_2}(\Xi) \qquad\qquad \square$$

The subspace $\mathbf{V}^0_h \subset \mathbf{V}_h$ h is the space of discrete functions that vanish on the boundary of Ω.

3.4. The Variational Formulation Discretization

We consider a classic mixed variational discretization of problem (4) in primitive variables [16,17], in which an approximation $(\vec{u}_h, p_h)$ to the exact solution $(\vec{u}, p)$ of (4) is obtained by solving the problem

$$
\begin{aligned}
a(\vec{u}_h, \vec{v}_h) + b(\vec{v}_h, p_h) &= 0 \quad \forall \vec{v}_h \in \mathbf{V}^0_h \\
b(\vec{u}_h, q_h) &= 0 \quad \forall q_h \in M_h,
\end{aligned}
\tag{16}
$$

where $(\mathbf{V}_h, M_h)$ are couples of finite-dimensional spaces parameterized with h, as introduced above, and with the bilinear forms a and b defined as

$$a\,(\vec{w}_h, \vec{v}_h) \;=\; \nu \int_\Omega \nabla \vec{w}_h : \nabla \vec{v}_h \, d\Omega + \frac{\nu}{k} \int_\Omega \vec{w}_h \cdot \vec{v}_h \, d\Omega$$

$$b\,(\vec{v}_h, q_h) \;=\; -\int_\Omega \nabla \cdot \vec{v}_h \, q_h \, d\Omega.$$

The conditions for the well posedness of a saddle point system is known as inf-sup conditions or Ladyzhenskaya–Babuška–Breezi (LBB) condition:

$$\underset{q \in M_h, q_h \neq 0}{\text{Inf}} \;\; \underset{\vec{v}_h \in \mathbf{V_h}, \vec{v}_h \neq \vec{0}}{\text{Sup}} \frac{b\,(\vec{v}_h, q_h)}{\|q_h\|_{L^2} \|\vec{v}_h\|_{H^1}} \geq C, \tag{17}$$

with C independent of the discritization parameter h, $\|\cdot\|_{L^2} \|\cdot\|_{H^1}$ are the classic norms defined over the spaces $L^2\,(\Omega)$ and $H^1\,(\Omega)$, respectively, as in [11]. The subspace choice M_h ensures the discrete velocity $\vec{u}_h$ solution of (16) is divergence-free.

In the context of IGA, we will use spline-based spaces $(\mathbf{V}_h^0, M_h)$, which satisfy the inf-sup condition (17). The elements considered can be seen as spline generalization of well-known finite elements spaces, namely Taylor–Hood elements [11]. Thanks to the high interelement regularity, which is the main feature of splines, the proposed discretizations are conforming, i.e., produce globally continuous, discrete velocities. Moreover, the spline generalization of Taylor–Hood elements also enjoys property [18] and thus provides divergence-free discrete solutions (15).

4. Numerical Results

In this section, we present some numerical experiments to predict the effect of fluid shear stress on cancer cells localized in the interstitial region to assess its metastatic behavior. For all cases, we use experimental data obtained in the works of [1,5–8].

We use GeoPDEs, an open source and free package for isogeometric analysis in Matlab [12,19], to achieved the numerical implementation of the discretized problem (16).

As a peculiarity of IGA is to allow for high degree and high regularity discretization spaces, most of the tests below are performed for degree $p = 5$ and regularity $\alpha = 4$.

We use multipatch geometries (Figure 4) and these discretizations are C^0 between the patches. When applied to incompressible flows, these discretizations produce pointwise divergence-free velocity fields and hence exactly satisfy mass conservation. We enforce the Dirichlet boundary conditions weakly by Nitsche's method, allowing this method to default to a compatible discretization of Darcy flow in the limit of vanishing viscosity [12].

Using the data from Table 1 we perform different experiments with different choices of region configuration and tumor size.

Table 1. Boundary conditions and parameters values.

Boundary Conditions		The Permeability k	The Viscosity ν
$\vec{u}_h = \vec{0}$ µs^{-1}	on Γ_T	3.5×10^{-3} [5]	5×10^{-6} [8]
$\vec{u}_h = (1,0)$ µs^{-1}	on Γ_I [20]		

4.1. Case 1

Let us begin by considering a simple circular case which will be used as a pattern for the others ones. We consider $\Omega = [-10, 10] \times [-10, 10]$ and a centred tumor region with a circular boundary, with radius $r = 2(\mu)$.

For this case we present the physical initial mesh, with control points, at Figure 5, an *h*-refinement (by multiple knot insertion) sequence, Figures 6 and 7, and the correspondent results for velocity and shear stress.

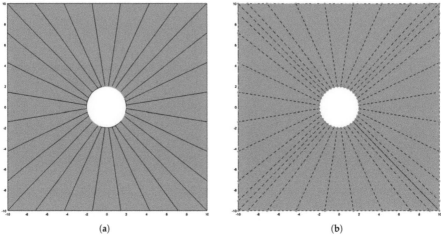

(a)   (b)

Figure 5. Case 1: physical domain, initial mesh and control points. (**a**) The physical mesh; (**b**) the control point for the NURBS surface.

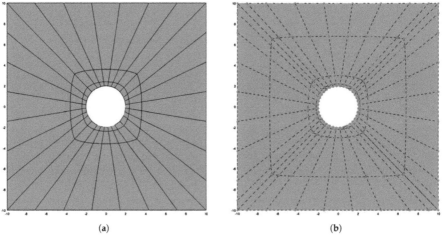

(a)   (b)

Figure 6. Case 1: physical domain, h_1-refinement mesh and correspondent control points. (**a**) The physical mesh; (**b**) the control point for the NURBS surface.

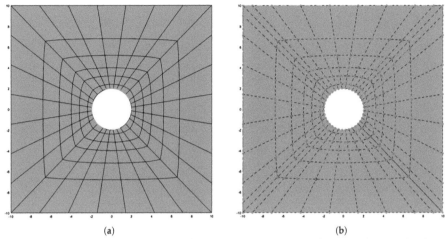

Figure 7. Case 1: physical domain, h_2-refinement mesh and correspondent control points. (**a**) The physical mesh; (**b**) the control point for the NURBS surface.

With this discretization the velocity magnitude and stream lines obtained is represented in Figure 8. The flow-generate shear stress is between -7×10^{-2} and 7×10^{-2} dyn/cm^2 (we recall 1 dyn/cm^2 = 0.1 Pa) over the whole domain, as shown in Figure 9, and maximum is achieved on the cell surface, this means on the membrane as expected.

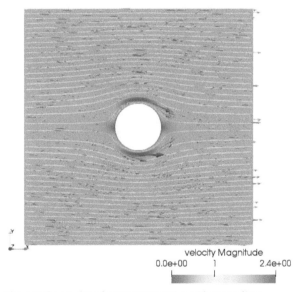

Figure 8. Case 1: the velocity representation with stream lines.

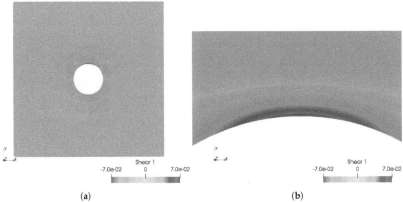

Shear 1
-7.0e-02   0   7.0e-02

Shear 1
-7.0e-02   0   7.0e-02

(a)   (b)

Figure 9. Case 1: the fluid shear stress representation, which is between -7×10^{-2} and 7×10^{-2} dyn/cm^2. (a) The fluid shear stress overview; (b) zoom over the higher shear stress zone.

Next, for this case, we will repeat the simulation with a *h*-refinement on the domain. First we consider at patches interfaces the new knots array $\Xi_{h_1} = [0.8, 0.95]$.

With this discretization the flow-generate shear stress is between -1.5 and 1.5 dyn/cm^2 over the whole domain, as shown in Figure 10, and velocity magnitude is between 0 and 2 μs^{-1}, Figure 11.

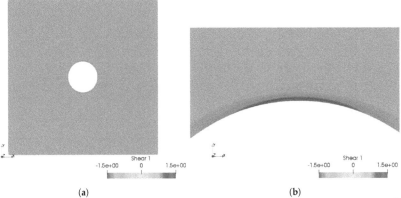

Shear 1
-1.5e+00   0   1.5e+00

Shear 1
-1.5e+00   0   1.5e+00

(a)   (b)

Figure 10. Case 1: the fluid shear stress representation, which is between -1.5 and 1.5 dyn/cm^2, for h_1-refinement. (a) The fluid shear stress overview; (b) zoom over the higher shear stress zone.

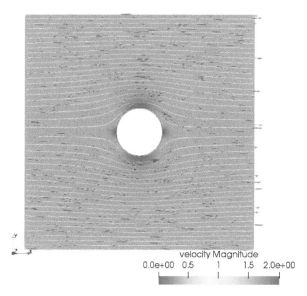

Figure 11. Case 1: the velocity representation with stream lines for domain h_1-refinement.

Continuing with the h-refinement, we consider at paches interfaces the new knots array $\Xi_{h_2} = [0.4, 0.6, 0.75, 0.85, 0.95]$.

With this discretization the flow-generate shear stress is between -1.5 and 1.5 dyn/cm^2 over the whole domain, as shown in Figure 12, and velocity magnitude is between 0 and 2 μs^{-1}, Figure 13, as in the previous refinement case.

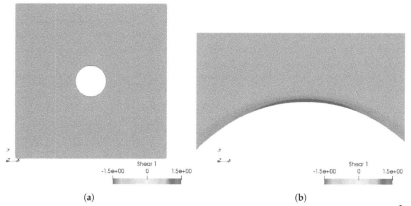

(a)   (b)

Figure 12. Case 1: the fluid shear stress representation, which is between -1.5 and 1.5 dyn/cm^2, for h_2-refinement. (**a**) The fluid shear stress overview; (**b**) zoom over the higher shear stress zone.

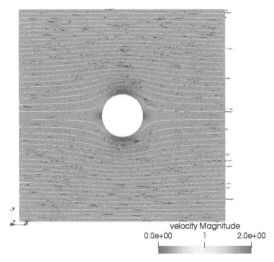

Figure 13. Case 1: the velocity representation with stream lines for domain h_2-refinement.

These numerical results show that mesh refinement does not affect results for $\mathcal{T}_{h_2}$. At this point, the model and its results are independent of the mesh.

Finally, we point to the velocity magnitude computed at any point $P \in \Omega$ using discrete ℓ_2 norm

$$\|\vec{u}(P)\| = \sqrt{u_x(P)^2 + u_y(P)^2}. \tag{18}$$

4.2. Case 2

Case 2 is similarly to case 1, we consider $\Omega = [-10, 10] \times [-10, 10]$ and a centred tumor region with a circular irregular boundary, as shown at Figure 14. Whereas the irregular circular line is obtained from perturbations on the line of the pattern case 1, we quantify the boundary irregularities by comparison between the cells' perimeter, introducing the parameter $\mathcal{I}$ (the irregularity parameter indexed with the case study number) defined as

$$\mathcal{I}_2 = \frac{|d_2 - d_1|}{d_1}. \tag{19}$$

with $d_1 = 4\pi = 12.5664$ and $d_2 = 13.3937$. For Case 2 we obtain $\mathcal{I}_2 = 0.0658$.

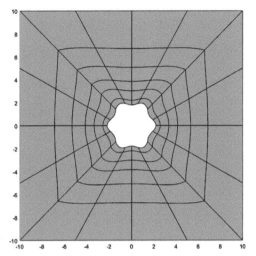

Figure 14. Case 2: physical domain and mesh with h_2-refinement.

With this cell configuration, $\mathcal{I}_2 = 0.0658$, the flow-generate shear stress is between -1.7 and 1.7 dyn/cm^2 over the whole domain, as shown in Figure 15, and velocity magnitude is between 0 and 2.2 µs^{-1}, Figure 16.

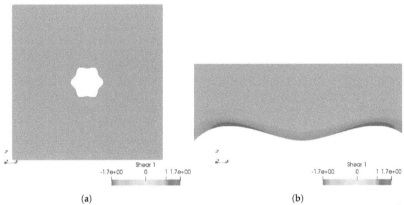

(a)   (b)

Figure 15. Case 2: the fluid shear stress representation, which is between -1.7 and 1.7 dyn/cm^2. (a) The fluid shear stress overview; (b) zoom over the higher shear stress zone.

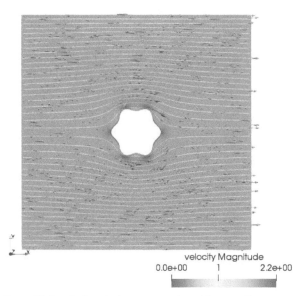

Figure 16. Case 2: the velocity representation with stream lines.

4.3. Case 3

As in case 2, we consider $\Omega = [-10, 10] \times [-10, 10]$ and a centred tumor region with a circular irregular boundary. Here we increase the irregularity parameter to $\mathcal{I}_3 = \frac{|13.8401 - 12.5664|}{12.5664} = 0.1014$, as shown in Figure 17.

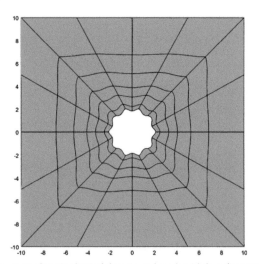

Figure 17. Case 3: physical domain and mesh with h_2-refinement.

With this cell configuration, $\mathcal{I}_3 = 0.1014$, the flow-generate shear stress is between -2.2 and 2.2 dyn/cm^2 over the whole domain, as shown in Figure 18, and velocity magnitude is between 0 and 2.5 μs^{-1}, Figure 19.

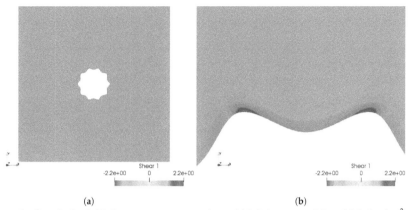

(a)   (b)

Figure 18. Case 3: the fluid shear stress representation, which is between -2.2 and 2.2 dyn/cm^2. (a) The fluid shear stress overview; (b) zoom over the higher shear stress zone.

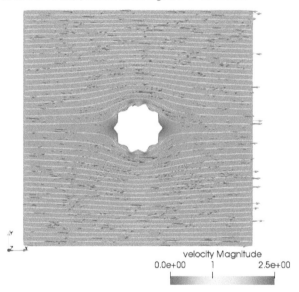

Figure 19. Case 3: the velocity representation with stream lines.

4.4. Case 4

Once again we consider $\Omega = [-10, 10] \times [-10, 10]$ and a centred tumor region with a circular irregular boundary and the irregularity parameter $\mathcal{I}_4 = \frac{|20.2692 - 12.5664|}{12.5664} = 0.613$, as shown at Figure 20.

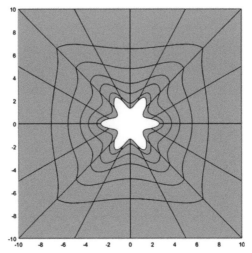

Figure 20. Case 4: physical domain and mesh with h_2-refinement.

With this cell configuration, $\mathcal{I}_4 = 0.613$, the flow-generate shear stress is between -3 and 3 dyn/cm^2 over the whole domain, as shown in Figure 21, and velocity magnitude is between 0 and 2.9 µs^{-1}, Figure 22.

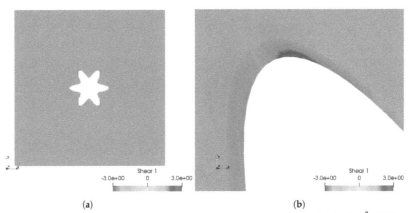

(a)   (b)

Figure 21. Case 4: the fluid shear stress representation, which is between -3 and 3 dyn/cm^2. (**a**) The fluid shear stress overview; (**b**) zoom over the higher shear stress zone.

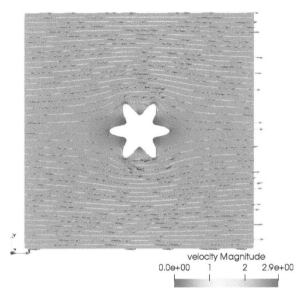

Figure 22. Case 4: the velocity representation with stream lines.

With the results from 1 to 4 we can define an evolution graphic in Figure 23 for shear stress maximum, the blue line, and velocity magnitude maximum, the red line. Regarding cell membrane resistance, at microenvironment context this evolution is quite important. We observed a linear dependence between the irregularity parameter and shear stress. The Figure 23 shows the dependence of shear stress on velocity and the irregularity parameter, given that the two lines have different growth rates.

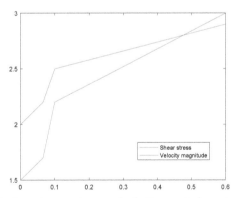

Figure 23. Evolution for shear stress maximum and velocity magnitude maximum by the irregularity parameter.

To conclude we present two more cases quite near to reality. Inspired by Figure 1 we define the outline shown in Figure 24. The following cases are determined from the outline 3D by cut planes for case 5, the green one, and case 6, the blue one.

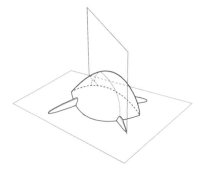

Figure 24. Tumor's 3D outline and cut planes for case 5, the green one, and case 6, the blue one.

4.5. Case 5

With this cell configuration, domain in Figure 25 (corresponding to the green section in Figure 24) we get $\mathcal{I}_5 = 0.613$, the flow-generate shear stress is between -2.8 and 2.5 dyn/cm^2 over the whole domain, as shown in Figure 26, and velocity magnitude is between 0 and 2.8 μs^{-1}, Figure 27. For this value of $\mathcal{I}_5$ we obtain the expected from comparison with the above cases. We point to the geometric asymmetric effect on the maximum value of shear stress.

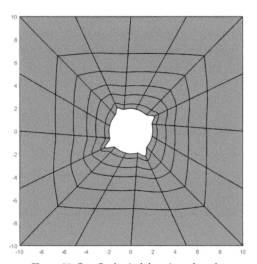

Figure 25. Case 5: physical domain and mesh.

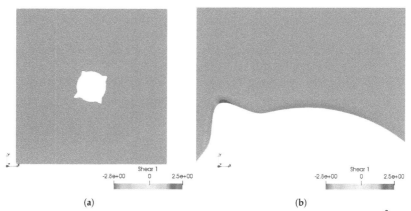

(a)   (b)

Figure 26. Case 5: the fluid shear stress representation, which is between -2.8 and 2.5 dyn/cm^2. (**a**) The fluid shear stress overview; (**b**) zoom over the higher shear stress zone.

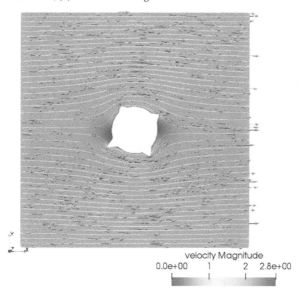

Figure 27. Case 5: the velocity representation with stream lines.

4.6. Case 6

Finally we consider for the cell configuration the blue section of the model in Figure 24, as shown in Figure 28. The flow-generate shear stress is between -4.1 and 3.3 dyn/cm^2 over the whole domain, as shown in Figure 29, and velocity magnitude is between 0 and 1.9 µs^{-1}, Figure 30.

With this result, we can see that the zone with the highest shear stress value is over the "body" of the cell and not in its branches. In fact, it is the branches we should focus on because they promote the displacement or duplication of the cell.

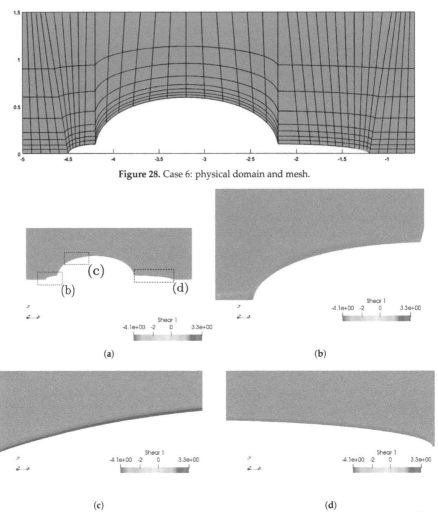

Figure 28. Case 6: physical domain and mesh.

(a)

(b)

(c)

(d)

Figure 29. Case 6: the fluid shear stress representation, which is between −4.1 and 3.3 dyn/cm². (**a**) The fluid shear stress overview; (**b**) zoom over the higher shear stress zone; (**c**) zoom over the higher shear stress zone; (**d**) zoom over the higher shear stress zone.

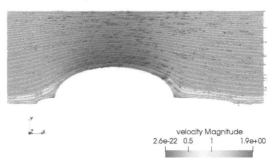

Figure 30. Case 6: the velocity representation with stream lines.

5. Conclusions

From this preliminary study, it is possible to establish the adequacy of the isogeometric analysis and NURBS to model irregular tumor cells and evaluate the shear forces at tumor microenvironments.

With some numerical examples, using different cell configurations, we have identified conditions that allow the increase or decrease of the fluid shear stress, which could contribute to the metastatic process by enhancing tumor cell invasion and circulating tumor cells. The cancer cell invasive potential is significantly reduced, as much as 92%, upon exposure to 0.55 dyn/cm^2 fluid shear stress [21]. This work contributed to establish a relationship between a quantification of cell membrane irregularity and the maximum value of shear stress evaluated on the cell membrane, by the effect of interstitial fluid in the microenvironment. In follow-up works, we must understand how the cell reacts to these forces.

Mathematical models of fluid shear stress effects coupled with in vitro and in vivo experimental validation, may better predict cell behavior in such dynamic microenvironments, and potentially provide novel approaches for the prevention of metastasis.

A particular feature of cancer modeling revolves around the idea that a tumors' size and shape changes over time and its resistance to the fluid shear stress is also affected by the presence of other substances in the interstitial region meaning, for this reason, is the motivation for further work.

Acknowledgments: The author acknowledges the support of IPL IDI and CA through the Project IGACFC 2018.

Conflicts of Interest: The author declares no conflict of interest.

References

1. Huang, Q.; Hu, X.; He, W.; Zhao, Y.; Hao, S.; Wu, Q.; Li, S.; Zhang, S.; Shi, M. Fluid shear stress and tumor metastasis. *Am. J. Cancer Res.* **2018**, *8*, 763–777. [PubMed]
2. Brooks, D.E. The biorheology of tumor cells. *Biorheology* **1984**, *21*, 85–91. [CrossRef] [PubMed]
3. Shieh, A.; Swartz, M. Regulation of tumor invasion by interstitial fluid flow. *Phys. Biol.* **2011**, *8*, 015012. [CrossRef] [PubMed]
4. Cottrell, J.; Hughes, T.; Bazilevs, Y. *Isogeometric Analysis: Toward Integration of CAD and FEA*; John Wiley & Sons: Hoboken, NJ, USA, 2009.
5. Chen, C.; Malkus, D.; Vanderby, R. A fiber matrix model for interstitial fluid flow and permeability in ligaments and tendons. *Biorheology* **1998**, *35*, 103–118. [CrossRef]
6. Hu, X.; Adamson, R.H.; Liu, B.; Curry, F.E.; Weinbaum, S. Starling forces that oppose filtration after tissue oncotic pressure is increased. *Am. J. Physiol. Heart Circ. Physiol.* **2000**, *279*, 1724–1736. [CrossRef] [PubMed]
7. Lartigue, J. Genomic Complexity Stifles Targeted Advances in Colorectal Cancer. *OncologyLive* **2016**, *17*. Available online: https://www.onclive.com/publications/Oncology-live/2016/Vol-17-No-4/genomic-complexity-stifles-targeted-advances-in-colorectal-cancer (accessed on 26 December 2019).
8. Ng, C.; Swartz, M. Fibroblast alignment under interstitial fluid flow using a novel 3-D tissue culture model. *Am. J. Physiol. Heart Circ. Physiol.* **2003**, *284*, 1771–1777. [CrossRef] [PubMed]
9. Butler, S.L.; Kohles, S.S.; Thielke, R.J.; Chen, C.; Vanderby, R., Jr. Interstitial fluid flow in tendons or ligaments: A porous medium finite element simulation. *Med. Biol. Eng. Comput.* **1997**, *35*, 742–746. [CrossRef] [PubMed]
10. Yao, W.; Ding, G.-H. Interstitial fluid flow: Simulation of mechanical environment of cells in the interosseous membrane. *Acta Mech. Sin.* **2011**, *27*, 602–610. [CrossRef]
11. Girault, V.; Raviart, P. *Finite Element Methods for Navier–Stokes Equations: Theory and Algorithms*; Springer Publishing Company, Incorporated: Berlin, Germany, 2011.
12. Falco, C.; Reali, A.; Vázquez, R. GeoPDEs: A research tool for Isogeometric Analysis of PDEs. *Adv. Eng. Softw.* **2008**, *42*, 1020–1034. [CrossRef]
13. Durlofsky, L.; Brady, J.F. Analysis of the Brinkman equation as a model for flow in porous media. *Phys. Fluids* **1987**, *30*, 3329–3341. [CrossRef]
14. Pnueli, D.; Gutfinger, C. *Fluid Mechanics*; Cambridge University Press: Cambridge, UK, 1992.
15. Piegl, L.; Tiller, W. *The NURBS Book*; Springer: Berlin, Germany, 1997.
16. Hosseini, B.; Möller, M.; Turek, S. Isogeometric Analysis of the Navier–Stokes equations with Taylor–Hood B-spline elements. *Appl. Math. Comput.* **2015**, *267*, 264–281. [CrossRef]

17. Buffa, A.; de Falco, C.; Sangalli, G. IsoGeometric Analysis: Stable elements for the 2D Stokes equation. *Int. J. Numer. Methods Fluids* **2011**, *65*, 1407–1422. [CrossRef]
18. Evans, J.; Hughes, T. Isogeometric divergence-conforming b-splines for the Darcy–Stokes–Brinkman equations. *Math. Models Methods Appl. Sci.* **2013**, *23*, 671–741. [CrossRef]
19. Vázquez, R. A new design for the implementation of isogeometric analysis in Octave and Matlab: GeoPDEs 3.0. *Comput. Math. Appl.* **2008**, *72*, 523–554. [CrossRef]
20. Dafni, H.; Israely, T.; Bhujwalla, Z.M.; Benjamin, L.E.; Neeman, M. Overexpression of vascular endothelial growth factor 165 drives peritumor interstitial convection and induces lymphatic drain. *Cancer Res.* **2002**, *62*, 6731–6739. [PubMed]
21. Shi, Z.; Tarbell, J. Fluid flow mechanotransduction in vascular smooth muscle cells and fibroblasts. *Ann. Biomed. Eng.* **2011**, *39*, 1608–1619. [CrossRef] [PubMed]

MDPI
St. Alban-Anlage 66
4052 Basel
Switzerland
Tel. +41 61 683 77 34
Fax +41 61 302 89 18
www.mdpi.com

Mathematical and Computational Applications Editorial Office
E-mail: mca@mdpi.com
www.mdpi.com/journal/mca

Lightning Source UK Ltd.
Milton Keynes UK
UKHW051148020920
369170UK00007B/134

9 783039 369522